# Mathematics Made Simple

**Fifth Edition**

## Abraham Sperling, Ph.D., and Monroe Stuart

**Revised by Christine M. Peckaitis**

Edited and prepared for publication by The Stonesong Press, Inc.

**A Made Simple Book**

**DOUBLEDAY** New York London Toronto Sydney Auckland

**Published by Doubleday**
a division of Bantam Doubleday Dell Publishing Group, Inc.
1540 Broadway, New York, New York 10036

**Made Simple** and **Doubleday** are trademarks of Doubleday,
a division of Bantam Doubleday Dell Publishing Group, Inc.

**Edited and prepared for publication by The Stonesong Press, Inc.**
**Managing Editor:** Sheree Bykofsky
**Design:** Blackbirch Graphics, Inc.
**Design Consultant:** Binns & Lubin/Martin Lubin
**Production Consultant:** RECAP: PUBLICATIONS, INC.

**Library of Congress Cataloging-in-Publication Data**
Sperling, Abraham Paul. 1912–
    Mathematics made simple/Abraham Sperling and
    Monroe Stuart.
    5th ed./revised by Christine M. Peckaitis.
        p.   cm.
    "A Made Simple book."
    1. Mathematics   I. Stuart, Monroe
II. Peckaitis, Christine M.   III. Title.
QA39.2.S684   1990        89-48249
512'1.–dc20                  CIP
ISBN 0-385-26584-0

JUNE 1991
FIFTH EDITION
10   9   8   7

# CONTENTS

# About This Book

**Mathematics Made Simple** was first published in the 1940s. It was designed especially for students in high school and college, for those who aspired to better jobs, or for those who desired to improve their mathematical skills.

This book serves as a review of arithmetic, and an introduction to algebra, geometry, and trigonometry. Combinations and permutations are covered carefully in the Probability chapter. The exercises and answers in this book provide readers with opportunities to test their mastery of each step in these common branches of mathematics.

Examples are given with small numbers so that the reader can concentrate on the principle rather than needlessly long explanations.

The weights and measures tables in the appendices are designed to be as useful as possible for practical applications. Among the tables in the appendices are the following: Table of Square Roots, Other Multiplication Tables, U.S. and Metric System Measures, and a Table of Trigonometric Functions.

## Features of This New Edition:

Each chapter has a **glossary** of mathematical terms introduced in that chapter. The glossary is useful as a study reference guide.

**Multiple Choice Tests** at the end of every chapter, and a final comprehensive test at the end of the book help monitor your progress. The tests should help you pinpoint possible areas for review. Consider 70 percent correct on a chapter test an average understanding of the chapter material.

**Practically Speaking** boxes introduce real life applications of mathematics in sections that teach the necessary principles.

**Answers** to all Exercise Sets, Tests, and Practically Speaking boxes are in the appendices.

For convenience, a list of **Common Mathematical Symbols** is included at the beginning of this book.

## A Note on the Use of Calculators

Because of their availability and relatively low cost, small hand-held calculators have replaced the laborious pencil-and-paper process of computation. Today, arithmetical and algebraic calculations in most offices and work places and in many classrooms are done with the aid of a calculator.

There are many models and types of calculators. Even calculators that are relatively simple and offer few special features have a distinctive design according to their manufacturer. It is worthwhile to become familiar with location of the keys so that calculations can be made quickly and correctly. The different arrangements from one model to another can lead to serious errors. Practice makes perfect, so one must practice because a calculator provides correct answers only when used as directed in the manual provided by the manufacturer. The right keys must be pressed in the proper order. A calculator's algebraic entry system allows a problem to be entered in the same order as it is written as in the four fundamental arithmetical calculations which follow:

*Addition*
123 + 456 = 579

*Subtraction*
789 − 456 = 333

*Multiplication*
12.3 × 4.5 = 55.35

*Division*
12.3 ÷ 4.5 = 2.7333333

Besides the basic arithmetical processes, a calculator can provide discounts, square roots, reciprocals, chain, and mixed calculations, all generally arrived at by touching the right key or keys. A mixed calculation example follows:

$$12 + (34 \times 56) - 789 = 1127$$

A student should remember that the aim is to understand and master the principles of the problem. The calculator is a helpful tool, not a crutch.

Although often appealing when seen in a store, a calculator that displays the time, the date, rings alarm bells, or has other special features is not essential to the understanding of decimals, fractions, powers, and roots.

## Common Mathematical Symbols

The following is a list of symbols frequently used in mathematics. Memorize this list if possible. Use this list for reference whenever a new symbol appears in this book.

| Symbol | Meaning of the Symbol in Words |
|---|---|
| > | is greater than |
| < | is less than |
| ≥ | is greater than or equal to |
| ≤ | is less than or equal to |
| ≅ | is congruent to |
| ≇ | is not congruent to |
| + | plus (addition) |
| − | minus (subtraction) |
| × | times (multiplication) |
| $ab$ | $a$ times $b$ |
| $a \times b$ | $a$ times $b$ |
| ÷ | divided by |
| $a \div b$ | $a$ divided by $b$ |

| | | | | |
|---|---|---|---|---|
| $\sqrt{x}$ | the positive square root of $x$ | | GCF | greatest common factor |
| $\sqrt[n]{x}$ | the $n$th root of $x$ | | GCD | greatest common divisor |
| $x^n$ | $x$ to the $n$th power | | $P(A)$ | probability of $A$ |
| $(x, y)$ | rectangular coordinates of a point in a plane | | $'$ | minutes |
| | | | $''$ | seconds |
| $m$ | slope | | $\{1, 2, 3\}$ | set of elements 1, 2, 3 |
| $b$ | $y$-intercept of a line | | $\{\ \}, 0$ | the empty set |
| $h$ | height of geometric figures | | $P(n,r)$ | permutation of $n$ things taken $r$ at a time |
| $b$ | base of geometric figures | | | |
| $l$ | length of geometric figures | | $!$ | factorial |
| $w$ | width of geometric figures | | $C(n,r)$ | combination of $n$ things $r$ at a time |
| $c$ | circumference | | | |
| $r$ | radius | | $\therefore$ | therefore |
| $d$ | diameter | | $\parallel$ | parallel |
| $\pi$ | pi, or approximately 3.14 | | $\perp$ | perpendicular |
| $|x|$ | absolute value of $x$ | | $\neq$ | not equal to |
| $\%$ | percent | | $\angle$ | angle |
| $a:b$ | the ratio of $a$ to $b$ | | $\circ$ | degree |
| $::$ | is in proportion to | | $\pm$ | plus or minus |
| LCD | least common denominator | | $\infty$ | infinity |

# Whole Numbers

## 1.1 Whole Numbers

**Arithmetic** is known as the science of numbers. We will learn how to work with whole numbers in this chapter. First, a different group of numbers will be defined.

The numbers 1, 2, 3, and so on are called **counting numbers.** They are called counting numbers because each counting number can be used to count physical objects. For example, we can label each apple in a bag with a counting number to find out how many apples we have.

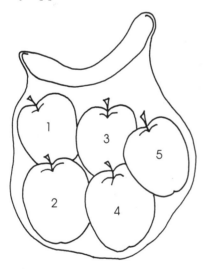

The counting numbers can be shown as the set {1,2,3 . . .}. The dots " . . ." mean "and so on."

We can count and add using the counting numbers. To subtract we may need the number zero. The **whole numbers** consist of the counting numbers together with the number zero. A whole number is a digit from 0 to 9, or a combination of digits, such as 17, 428, or 1,521. The set of whole numbers can be shown as {0,1,2,3 . . .}.

## 1.2 Addition of Whole Numbers

**Addition** is the process of finding the sum of two or more numbers. The numbers that are added together are called **addends.**

EXAMPLE 1: Add 2 + 4.

SOLUTION: 2 + 4 = 6

Note that if we add 2 + 4 the sum is 6. If we add 4 + 2 the answer is also 6. Numbers may be added up in any order and the sum will be exactly the same. This property of addition is known formally as the **Commutative Law of Addition.**

---

**Commutative Law of Addition**

$a + b = b + a$

---

The Commutative Law of Addition tells us that we can add two numbers in any order and still get the same answer. The numbers added on either side of the equals sign are exactly the same. The only difference is the order the numbers are added in.

EXAMPLE 2: Does 7 + 5 equal 5 + 7?

SOLUTION: If we apply the Commutative Law of Addition, we see that 7 + 5 equals 5 + 7. The only difference between the expressions 7 + 5 and 5 + 7 is the order of the numbers being added.

$$7 + 5 = 5 + 7$$
$$12 = 12$$

When three or more numbers are added, two of them are usually added first. Parentheses can be used to group together the numbers that are added together first. In EXAMPLE 3, add together 17 and 11 first.

EXAMPLE 3: Add (17 + 11) + 8.

SOLUTION: 
$$(17 + 11) + 8 =$$
$$28 + 8 =$$
$$36 =$$

Note what happens if we add 17 + (11 + 8)

$$17 + (11 + 8) =$$
$$17 + 19 =$$
$$36 =$$

The answer is still 36. This fact illustrates the **Associative Law of Addition.**

---

**Associative Law of Addition**

$$(a + b) + c = a + (b + c)$$

---

The Associative Law of Addition tells us that the way numbers are grouped does not affect the sum. Look at EXAMPLE 3 again. Add both ways to check that the sums are exactly the same. Using parentheses to group the numbers will not change the sum.

We also need to learn how to add together columns of numbers. One way to acquire speed in column addition is to learn to group successive numbers at sight and form larger numbers. Learn first to recognize groups of numbers that add up to 10.

EXAMPLE 4: Add the column of numbers. Use the hint on how to group if necessary.

```
  6,737
  7,726 ⎫
  2,884 ⎭ 10
  8,825
  2,201 ⎫
  4,669 ⎭ 10
  1,608
+ 2,599
```

SOLUTION: The sum of the column of figures is 37,249.

We need not limit ourselves to groups of only two numbers. Learn to combine three or more numbers together by sight to add up to 10.

EXAMPLE 5: Add the column of numbers. Use the hint to group together numbers that add up to 10.

$$5,511$$
$$5,522$$
$$8,113$$
$$2,037$$
$$8,474$$
$$7,745$$
$$\left.\begin{array}{r}3,355\\ +\ 4,505\end{array}\right\}10$$

SOLUTION: The sum of the column of figures is 45,262.

## Exercise Set 1.2

Use the Commutative Law of Addition to decide whether the following equations are true or false.

1. $2 + 29 = 25 + 2$
2. $75 + 8 = 75 - 8$
3. $16 + 16 = 16 + 16$
4. $128 + 41 = 4 + 128$
5. $11 + 3 = 3 + 11$

Use the Associative Law of Addition to decide whether the following equations are true or false.

6. $(29 + 14) + 3 = 29 + (14 + 3)$
7. $45 + (16 + 110) = (4 + 5 + 110)$
8. $(56 + 33) + 13 = 56 + (33 + 13)$
9. $44 + (11 + 81) = (44 + 11) + 81$
10. $34 + (25 + 45) = (34 + 25) + 60$

Find the sums by adding the numbers together mentally, if possible.

11. $53 + 13 =$
12. $64 + 28 =$
13. $59 + 17 =$
14. $65 + 38 =$
15. $118 + 48 =$
16. $139 + 46 =$
17. $178 + 57 =$
18. $274 + 89 =$
19. $457 + 76 =$
20. $326 + 134 =$
21. $495 + 179 =$
22. $697 + 267 =$

23. $673 + 568 =$
24. $878 + 595 =$
25. $1,578 + 673 =$
26. $11 + 4 =$
27. $15 + 3 =$
28. $13 + 6 =$
29. $23 + 5 =$
30. $25 + 8 =$
31. $35 + 6 =$
32. $64 + 9 =$
33. $19 + 18 =$
34. $13 + 19 =$
35. $32 + 29 =$
36. $63 + 16 =$
37. $54 + 38 =$
38. $69 + 27 =$
39. $75 + 38 =$
40. $118 + 58 =$
41. $149 + 36 =$
42. $178 + 67 =$
43. $264 + 79 =$
44. $467 + 66 =$
45. $336 + 144 =$
46. $479 + 195 =$
47. $687 + 257 =$
48. $693 + 578 =$
49. $888 + 585 =$
50. $1,468 + 724 =$
51. $13 + 5 =$
52. $35 + 3 =$
53. $43 + 4 =$
54. $52 + 6 =$
55. $35 + 7 =$
56. $47 + 7 =$
57. $74 + 9 =$
58. $21 + 28 =$
59. $15 + 17 =$
60. $42 + 39 =$
61. $63 + 18 =$
62. $64 + 38 =$
63. $79 + 27 =$
64. $85 + 48 =$
65. $116 + 38 =$
66. $139 + 46 =$
67. $168 + 47 =$
68. $254 + 89 =$
69. $346 + 74 =$
70. $457 + 134 =$
71. $579 + 115 =$
72. $677 + 237 =$
73. $683 + 568 =$
74. $878 + 595 =$
75. $1,558 + 723 =$

Add each column of numbers. Look for groupings of 10 as you add the columns to help find a total.

76.
$$\begin{array}{r}67\\ \left.\begin{array}{r}28\\ 22\end{array}\right\}10\\ 14\\ 55\\ 82\\ 87\\ +\ 34\end{array}$$

77.
$$\begin{array}{r}524\\ 616\\ 546\\ 534\\ 824\\ 377\\ 882\\ +\ 665\end{array}$$

| **78.** | 551 | **79.** | 2,642 |
|---|---|---|---|
| | 473 | | 6,328 |
| | 572 | | 2,060 |
| | 468 | | 9,121 |
| | 246 | | 3,745 |
| | 455 | | 5,545 |
| | 264 | | 6,474 |
| | + 455 | | + 5,567 |

| **80.** | 28 |
|---|---|
| | 76 |
| | 88 |
| | 27 |
| | 54 |
| | 21 |
| | 85 |
| | + 69 |

## 1.3 Subtraction of Whole Numbers

**Subtraction** is the process of finding the difference between two numbers. This is the same as finding out how much must be added to one number, called the **subtrahend,** to equal another, called the **minuend.** Use the minus sign (−) to indicate subtraction.

For example, when we subtract 12 from 37, the difference is 25. The difference, 25, plus the subtrahend, 12, must equal the minuend, 37. For this reason, addition is a good way to check subtraction.

There are several ways subtraction can be indicated verbally.

EXAMPLE 1: Use words to name four different ways of saying 16 − 4.

SOLUTION:

1. Subtract four from sixteen.
2. How much less than sixteen is four?
3. What is the difference between four and sixteen?
4. How much more than four is sixteen?

Each of the four solutions above means subtract 4 from 16.

EXAMPLE 2: Subtract 4 from 16.

SOLUTION: 16 − 4 = 12

Note that the answer to each of the verbal statements in EXAMPLE 1 is 12, after we translate the verbal statements into numerical expressions.

When we subtract two large numbers, and we are not using a calculator for the computation, we usually stack the two numbers to make our calculation easier to perform. When we stack the two numbers, we are careful to align numbers with the same place value. We will use the borrowing method to subtract two numbers.

EXAMPLE 3: Subtract 9,624 − 5,846.

SOLUTION: First, stack the numbers so that the larger number is on top.

Then begin subtracting. Since subtracting 6 from 4 does not make any sense when working with whole numbers, we borrow a ten from the tens column. We cross out the 2 in the tens column, subtract 1 ten from the 2 tens, and write 1 above the crossed-out 2. The 4 becomes 14, or 4 ones and 1 ten, so we can continue with our subtraction. Subtracting 6 from 14 equals 8.

$$\begin{array}{r} {}^{1}\ {}^{14} \\ 9{,}6\,\not2\,\not4 \\ -\,5{,}8\,4\,6 \\ \hline 8 \end{array}$$

We may continue subtracting by using the borrowing method. Read the rest of the subtraction as: 4 from 11 leaves 7, 8 from 15 leaves 7, 5 from 8 leaves 3.

$$\begin{array}{r} {}^{8}\ {}^{5}\ {}^{1} \\ 9,\ 6\ 2\ 4 \\ 5,\ 8\ 4\ 6 \\ \hline 3,\ 7\ 7\ 8 \end{array}$$

The difference between 5,846 and 9,624 is 3,778.

## Exercise Set 1.3

Subtract to find the difference.

1. $123 - 56 =$
2. $67 - 24 =$
3. $345 - 12 =$
4. $77 - 34 =$
5. $237 - 237 =$
6. $450 - 422 =$
7. $110 - 55 =$
8. $89 - 49 =$
9. $16 - 12 =$
10. $441 - 250 =$

Use the borrowing method to perform these subtractions.

11. $956,224 - 23,478 =$
12. $341,288,543 - 12,347,632 =$
13. $588,257,964 - 412,973,833 =$
14. $23,145,742,789 - 23,145,634,500 =$
15. $387,429,533,756 - 231,634,677 =$

## 1.4 Multiplication of Whole Numbers

**Multiplication** is a short method of adding a number to itself a given number of times. The given number is called the **multiplicand.** The number of times the number is to be added is called the **multiplier.** The result is called the **product.** The multiplication sign ($\times$) indicates multiplication. When reading a multiplication problem aloud, read "times" for $\times$.

For instance, 4 times 15 means $15 + 15 + 15 + 15$. Fifteen is added four times.

EXAMPLE 1: Multiply 50 by 3.

SOLUTION: Here 50 is the multiplicand, and 3 is the multiplier. To find the product, we can add 50 to itself three times.

$$\begin{array}{r} 50 \text{ (multiplicand)} \\ \times\ 3 \text{ (multiplier)} \\ \hline 150 \text{ (product)} \end{array}$$

Note that $50 \times 3$ is the same multiplication problem as 50
$\times\ 3$

**MULTIPLICATION TABLE FROM TWO TO TWELVE**

| NO. | ×2 | ×3 | ×4 | ×5 | ×6 | ×7 | ×8 | ×9 | ×10 | ×11 | ×12 |
|---|---|---|---|---|---|---|---|---|---|---|---|
| 1 | 2 | 3 | 4 | 5 | 6 | 7 | 8 | 9 | 10 | 11 | 12 |
| 2 | 4 | 6 | 8 | 10 | 12 | 14 | 16 | 18 | 20 | 22 | 24 |
| 3 | 6 | 9 | 12 | 15 | 18 | 21 | 24 | 27 | 30 | 33 | 36 |
| 4 | 8 | 12 | 16 | 20 | 24 | 28 | 32 | 36 | 40 | 44 | 48 |
| 5 | 10 | 15 | 20 | 25 | 30 | 35 | 40 | 45 | 50 | 55 | 60 |
| 6 | 12 | 18 | 24 | 30 | 36 | 42 | 48 | 54 | 60 | 66 | 72 |
| 7 | 14 | 21 | 28 | 35 | 42 | 49 | 56 | 63 | 70 | 77 | 84 |
| 8 | 16 | 24 | 32 | 40 | 48 | 56 | 64 | 72 | 80 | 88 | 96 |
| 9 | 18 | 27 | 36 | 45 | 54 | 63 | 72 | 81 | 90 | 99 | 108 |
| 10 | 20 | 30 | 40 | 50 | 60 | 70 | 80 | 90 | 100 | 110 | 120 |
| 11 | 22 | 33 | 44 | 55 | 66 | 77 | 88 | 99 | 110 | 121 | 132 |
| 12 | 24 | 36 | 48 | 60 | 72 | 84 | 96 | 108 | 120 | 132 | 144 |

To solve problems by multiplying, we must be able to compute the product. Here is the multiplication table from 2 to 12. We can use a calculator to check each product. Memorizing this table will help us solve problems more quickly, and will also reduce the chances of making a careless error.

When we multiply two numbers, we find that the Commutative Law can be applied to multiplication as well as to addition.

---

**Commutative Law of Multiplication**

$a \times b = b \times a$

---

The **Commutative Law of Multiplication** shows that when we multiply two numbers, the result is the same no matter what order we multiply them in.

In this next example, we see how to multiply more than two numbers together.

EXAMPLE 2: Multiply $(2 \times 30) \times 16$.

SOLUTION: $(2 \times 30) \times 16 =$
$60 \times 16 =$
$960 =$

Note what happens if we group the numbers together differently and multiply them together.

$2 \times (30 \times 16) =$
$2 \times 480 =$
$960 =$

The answer is the same no matter in what order the numbers are multiplied together. This is stated formally in the **Associative Law of Multiplication.**

---

**Associative Law of Multiplication**

$a \times (b \times c) = (a \times b) \times c$

---

**Estimation** is a way of guessing the approximate answer to a question. Estimation allows us to approximate an answer quickly instead of working it out exactly.

When we estimate in multiplication, we can round the multiplier and the multiplicand. Round to the nearest five, ten, or hundred depending on what makes sense when answering the question. If we round to the nearest ten, then we will round up numbers that have a value of 5 or more in the ones place.

EXAMPLE 3: Round 41, 37, 32, and 16 to the nearest ten.

SOLUTION: The numbers round up when the value in the ones place is closer to 10 than to 0. The numbers round down when the value in the ones place is between 1 and 4.

41 rounds down to 40

37 rounds up to 40

32 rounds down to 30

16 rounds up to 20

EXAMPLE 4: Round the numbers 273, 82, 144, and 641 to the nearest hundred.

SOLUTION: The numbers round up when the value in the tens place is between 5 and 9. Pay no attention to the values in the ones place, as the ones will not affect the answer. The numbers round down when the value in the tens place is between 1 and 4.

273 rounds up to 300

82 rounds up to 100

144 rounds down to 100

641 rounds down to 600

EXAMPLE 5: How many prizes have the 234 members of the Lions Club won together at the carnival, if each member has won 16 prizes? Find an estimated answer to the nearest hundred.

SOLUTION: To get an exact answer to this question, we multiply 234 by 16. To estimate this answer, try first rounding 234 to the nearest ten:

234 rounds down to 230

Then round 16 to the nearest ten:

16 rounds up to 20

Now multiply the new multiplier and multiplicand:

```
  230
× 20
4,600
```

Now multiply the original problem to see how close the estimate is.

```
  234
× 16
3,744
```

Is the rounded answer estimated within a hundred of the actual answer? No, the estimated answer is nearly 1,000 more than the actual answer. How can we get a closer answer to the exact answer by estimating? We can try multiplying the rounded number of members by the exact number of prizes won.

```
  230
× 16
3,680
```

Note that this answer is within a hundred of the exact answer.

---

**PRACTICALLY SPEAKING 1.4**

Jack wants to buy 12 CD's that cost $14 each.

1. How much money does Jack need to buy the 12 CD's?
2. If Jack has $85, how many CD's can he buy?

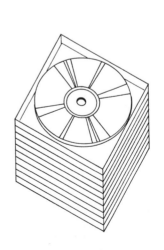

See Appendix F for the answers.

## Exercise Set 1.4

Find an equivalent expression by using the Commutative Law of Multiplication.

**1.** $32 \times 47 =$          **4.** $217 \times 21 =$

**2.** $123 \times 43 =$          **5.** $136 \times 24 =$

**3.** $182 \times 52 =$

Multiply, using the Associative Law of Multiplication to simplify the problems.

**6.** $1{,}112 \times (893 \times 14) =$

**7.** $(4{,}457 \times 369) \times 48 =$

**8.** $(48 \times 48) \times 61 =$

**9.** $83 \times (52 \times 73) =$

**10.** $115 \times (112 \times 46) =$

Multiply.

**11.** $83 \times 53 =$          **14.** $87 \times 87 =$

**12.** $115 \times 115 =$          **15.** $96 \times 46 =$

**13.** $4{,}562 \times 1{,}211 =$

Round each number to the place indicated.

**16.** Round 24,644 to the nearest hundred.

**17.** Round 978 to the nearest ten.

**18.** Round 14 to the nearest ten.

**19.** Round 4,932 to the nearest thousand.

**20.** Round 757 to the nearest hundred.

Estimate each answer by rounding both the multiplicand to the nearest hundred, and the multiplier to the nearest ten.

**21.** $693 \times 71 =$          **24.** $5{,}487 \times 378 =$

**22.** $451 \times 29 =$          **25.** $2{,}964 \times 833 =$

**23.** $3{,}981 \times 240 =$

## 1.5 Division of Whole Numbers

**Division** is the process of finding out how many times one number, called the **divi-sor,** goes into another number, called the **dividend.** The answer is called the **quotient.**

The sign for division is ÷ and this is read "divided by."

Division can also be described as the reverse of multiplication. This is true because in division, you undo the operations performed in multiplication.

For instance, since $15 \times 4 = 60$, the dividend 60 divided by the divisor 4 produces the quotient 15. Write this as:

$$60 \div 4 = 15$$

or as:

$$\text{divisor} \rightarrow 4\overline{)60} \quad \substack{15 \leftarrow \text{quotient} \\ \ \leftarrow \text{dividend}}$$

EXAMPLE 1: Divide 20 by 5.

SOLUTION: $20 \div 5 = 4$

When a divisor does not divide into a dividend an exact number of times, the number left over is called the **remainder.**

EXAMPLE 2: Divide 63 by 4.

SOLUTION: $63 \div 4 = 15$ with a remainder of 3

When writing the answer to a division problem, we can write "R" instead of "remainder." That is, if we were to write out the answer to EXAMPLE 2, we could write "15 R3" instead of "15 with a remainder of 3."

To check our answers to division problems, multiply the divisor by the quotient and add the remainder, if there is one. If our answer is correct, we will end up with our original dividend.

Let's check the answer to EXAMPLE 2. First, multiply the divisor by the quotient:

$$\begin{array}{r} 4 \text{ divisor} \\ \times 15 \text{ quotient} \\ \hline 60 \text{ new product} \end{array}$$

Since there is a remainder of 3, add it to the new product.

$$\begin{array}{r} 60 \text{ new product} \\ + \ 3 \text{ remainder} \\ \hline 63 \end{array}$$

Does this final sum match the original dividend in EXAMPLE 2? The original dividend was 63, so the answers match.

The method of the above examples is called short division because the intermediate steps can be worked out mentally. Another method, called long division, is exactly the same, but its intermediate steps are written out, as in the next example.

EXAMPLE 3: Divide 127,229 by 456.

SOLUTION: Set up the division so that the divisor is to the left of the dividend, and the dividend is under the $\overline{)}$. The $\overline{)}$ also means division.

```
                279    (quotient)
(divisor)  456)127,229  (dividend)
               912
               3602
               3192
               4109
               4104
                  5    (remainder)
```

The last digits, 2 and 9, of the dividend in such an example are "brought down" in the intermediate steps.

The quotient is 279 R5.

## Checking Answers

In general, additions are checked by adding the addends in a different order.

Subtraction is checked by adding the subtrahend to the remainder. The sum should equal the minuend. In other words, in a stacked subtraction example, the sum of the middle and bottom numbers should equal the top number.

Simple multiplication may be checked by reversing the multiplicand and multiplier and multiplying again.

Simple division may be checked by multiplying the divisor by the quotient, and then adding the remainder, if there is one.

## Computing Averages

To find the average of several quantities, divide their sum by the number of quantities.

EXAMPLE 4: What was the average attendance of people at a church if the daily attendance from Monday through Friday was as follows: 462, 548, 675, 319, and 521?

SOLUTION: First add the quantities together. Then divide the sum by the number of days, or 5.

$$\begin{array}{r} 462 \\ 548 \\ 675 \\ 319 \\ + \ 521 \\ \hline 2,525 \end{array}$$

$$2,525 \div 5 = 505$$

Our quotient is the average, so the average attendance at the church is 505 people.

**PRACTICALLY SPEAKING 1.5**

Alice is making up a monthly budget for herself. She has kept records for the past 8 months on the amount of money she spends.

| | |
|---|---|
| APRIL | $800 |
| MAY | $670 |
| JUNE | $429 |
| JULY | $620 |
| AUGUST | $520 |
| SEPTEMBER | $740 |
| OCTOBER | $650 |
| NOVEMBER | $780 |

1. How much more did Alice spend in April than in August?
2. How much less did Alice spend in July than in November?
3. What is the average amount she spent each month, based on the 8 months she kept records?

See Appendix F for the answers.

## Exercise Set 1.5

Divide, and then check your answers.

1. $7,258 \div 19 =$
2. $13,440 \div 35 =$
3. $21,492 \div 53 =$
4. $19,758 \div 37 =$
5. $47,085 \div 73 =$
6. $45,522 \div 54 =$
7. $42,201 \div 46 =$
8. $66,822 \div 74 =$
9. $53,963 \div 91 =$
10. $25,543 \div 16 =$

Divide using the long division method.

11. $47,974 \div 83 =$
12. $21,954 \div 67 =$
13. $88,445 \div 95 =$
14. $90,100 \div 123 =$
15. $229,554 \div 234 =$
16. $307,050 \div 345 =$
17. $59,448 \div 96 =$
18. $66,994 \div 86 =$
19. $47,320 \div 52 =$
20. $45,414 \div 62 =$
21. $78,027 \div 93 =$
22. $31,806 \div 38 =$

## Chapter 1 Glossary

**Addends**   The numbers being added together to make a sum.

**Addition**   The process of finding the sum of two or more numbers.

**Arithmetic**   The science of numbers.

**Associative Law of Addition**
$(a + b) + c = a + (b + c)$

**Associative Law of Multiplication**
$a \times (b \times c) = (a \times b) \times c$

**Commutative Law of Addition**
$a + b = b + a$
**Commutative Law of Multiplication**
$a \times b = b \times a$
**Counting Numbers** The set $\{1, 2, 3 \ldots\}$.
**Dividend** The number that is divided by the divisor.
**Division** The process of finding out how many times the divisor goes into the dividend.
**Divisor** The number that divides the dividend.
**Estimation** A way of approximating the answer to a question quickly.
**Minuend** The number the subtrahend is subtracting from.
**Multiplicand** The given number being multiplied.
**Multiplication** A short method of adding a number to itself a given number of times.
**Multiplier** The number of times the multiplicand is to be multiplied.
**Product** The result of the multiplication of the multiplicand and the multiplier.
**Quotient** The result of the division of the dividend by the divisor.
**Remainder** The amount left over after a division when the divisor does not divide into a dividend exactly.
**Subtraction** The process of finding the difference between two numbers.
**Subtrahend** The number subtracted from the minuend.
**Whole Numbers** The set $\{0, 1, 2, 3 \ldots\}$.

# Chapter 1 Test

For each problem, five answers are given. Only one answer is correct. After you solve each problem, check the answer that agrees with your solution.

1. A dealer bought 3 loads of coal weighing 6,242 pounds, 28,394 pounds, and 143,686 pounds. How much coal did he buy in all?

   A) 76,324 ___   D) 165,432 ___
   B) 178,322 ___   E) 376,178 ___
   C) 268,422 ___

2. If you earn $152 a week, how much will you earn in 12 weeks?

   A) $1,800 ___   D) $1,956 ___
   B) $1,884 ___   E) $1,742 ___
   C) $1,824 ___

3. Karen and Steve hiked 48 miles in 5 days. The first day they hiked 12 miles, the second day 9 miles, the third day 7 miles, the fourth day 9 miles. How many miles did they hike on the last day?

   A) 11 ___   D) 20 ___
   B) 8 ___   E) 14 ___
   C) 16 ___

4. How many packs of gum can you buy for $3.00 at the rate of 2 for 60 cents?

   A) 5 ___   D) 20 ___
   B) 10 ___   E) 30 ___
   C) 15 ___

5. If an automobile travels 450 yards in 15 seconds, how many feet does it go in $\frac{1}{3}$ of a second?

   A) 30 ___   D) 10 ___
   B) 90 ___   E) 50 ___
   C) 60 ___

**6.** An airplane hangar is 100 feet long, 50 feet wide, and 10 feet high. Estimate the cost of heating it at the rate of $25 per 1,000 cubic feet per season.

A) $125 ___  
B) $250 ___  
C) $1,250 ___  
D) $2,250 ___  
E) $2,500 ___

**7.** It takes 5 pounds of cement to cover 10 square feet. How many pounds of cement will be needed to cover a rectangular area 25 feet by 10 feet?

A) 25 ___  
B) 150 ___  
C) 200 ___  
D) 125 ___  
E) 130 ___

(*Note:* In solving problems such as No. 7 always *determine first what one unit will do.* In this case:
   If 5 pounds cover 10 square feet, then 1 pound covers 2 square feet.)

**8.** Mr. Curran sold 22 acres of his 142-acre estate to Mr. Brown, 30 acres to Mr. Jones, 14 acres to Mr. Smith, and 16 acres to Ms. Ives. How many acres did he have left?

A) 30 ___  
B) 40 ___  
C) 50 ___  
D) 60 ___  
E) 70 ___

**9.** Two machinists operating the same lathe work 10 hours each on a day- and on a night-shift respectively. One man turns out 400 pieces an hour, the other 600 pieces per hour. What will be the difference in their output at the end of 30 days?

A) 10,000 ___  
B) 6,000 ___  
C) 60,000 ___  
D) 40,000 ___  
E) 80,000 ___

**10.** Find the sum of 632 and 421.

A) 963 ___  
B) 1,161 ___  
C) 1,053 ___  
D) 1,543 ___  
E) 1,061 ___

**11.** Find the difference between 7,265 and 915.

A) 8,180 ___  
B) 6,450 ___  
C) 8,350 ___  
D) 6,180 ___  
E) 6,350 ___

**12.** Find the product of 63 and 17.

A) 46 ___  
B) 1,071 ___  
C) 1,270 ___  
D) 80 ___  
E) 1,520 ___

**13.** Divide 231 by 3.

A) 77 ___  
B) 16 ___  
C) 11 ___  
D) 89 ___  
E) 69 ___

**14.** Multiply 220 by 11.

A) 2,200 ___  
B) 2,450 ___  
C) 2,420 ___  
D) 2,222 ___  
E) 2,440 ___

**15.** Divide 287 by 41.

A) 11 ___  
B) 7 ___  
C) 16 ___  
D) 13 ___  
E) 4 ___

**16.** Divide 3,115 by 35.

A) 98 ___  
B) 77 ___  
C) 67 ___  
D) 89 ___  
E) 43 ___

# Fractions

## 2.1 Fractions

Though the product of any two whole numbers is always another whole number, the quotient of two whole numbers may or may not be a whole number. For instance, $2 \times 3 = 6$, and $6 \div 2 = 3$ but $2 \div 3$ does not equal a whole number. This sort of quotient is called a fractional number or a **fraction.**

More precisely, a **fraction** is an expression where the dividend, called the fraction's **numerator,** is written over the divisor, called the fraction's **denominator,** with a slanting or horizontal line between them to indicate the intended division. Thus, in common fraction form:

$$2 \div 3 = 2/3 \text{ or } \frac{2}{3}$$

with 2 as the numerator over 3 as the denominator.

From this we see that $\frac{2}{3}$ by definition means $2 \div 3$, or "2 divided by 3." Likewise, $\frac{3}{2}$ by definition means $3 \div 2$, or "3 divided by 2." However, we note that arithmetically:

$$\frac{2}{3} = 2 \times \frac{1}{3}, \text{ and } \frac{3}{2} = 3 \times \frac{1}{2}.$$

We read the symbol "$\frac{2}{3}$" as "two thirds," and the symbol "$\frac{3}{2}$" as "3 halves."

A **proper fraction** has a value less than 1 because, by definition, it has a numerator smaller than its denominator. Examples are $\frac{2}{3}$, $\frac{1}{4}$, and $\frac{3}{5}$.

An **improper fraction** has a value greater than 1 because, by definition, it has a numerator larger than its denominator. Examples: $\frac{3}{2}$, $\frac{7}{4}$, and $\frac{31}{9}$. But it is quite "proper" arithmetically to treat these fractions just like other fractions.

A **mixed number** consists of a whole number and a fraction written together with the understanding that they are to be added to one another. Examples are $1\frac{3}{4}$, which means $1 + \frac{3}{4}$, and $2\frac{5}{7}$, which means $2 + \frac{5}{7}$.

A **simple fraction** is one in which both numerator and denominator are whole numbers.

A **complex fraction** is one in which either the numerator or the denominator is a fraction or a mixed number, or in which both the numerator and the denominator are fractions or mixed numbers. Examples are:

$$\frac{\frac{1}{4}}{2}, \quad \frac{2}{\frac{2}{4}}, \quad \frac{\frac{1}{4}}{\frac{1}{3}}, \quad \frac{1\frac{1}{2}}{2}, \quad \frac{2}{3\frac{1}{4}}, \quad \text{and} \quad \frac{1\frac{1}{2}}{2\frac{3}{4}}.$$

In much of our work with fractions we need to apply a property called the **Fundamental Property for Fractions.** The property stated in words says that when the numerator and denominator of a fraction are both multiplied and divided by the same rational expression, the value of the fraction stays the same. The expression must not equal zero, however.

---

**Fundamental Property for Fractions**

If $\frac{a}{b}$ is a fraction, and $C$ is any rational expression not equal to zero, then

$$\frac{AC}{BC} = \frac{A}{B}$$

---

EXAMPLE 1: Find two fractions equal to $\frac{1}{2}$.

SOLUTION: We multiply both the numerator and the denominator of $\frac{1}{2}$ by any number to get an equivalent fraction.

$$\frac{1}{2} = \frac{1 \times (2)}{2 \times (2)} = \frac{2}{4} = \frac{2 \times (25)}{4 \times (25)} = \frac{50}{100}$$

From this example, we see that any common fraction can be written in as many different forms as we want, provided that we always multiply both the numerator and the denominator by the same expression.

A fraction written with the smallest possible whole numbers for both its numerator and its denominator is called a fraction in its **lowest terms.** Thus, of the fractions $\frac{1}{2}$, $\frac{2}{4}$, and $\frac{50}{100}$ in EXAMPLE 1, only $\frac{1}{2}$ is a fraction in lowest terms.

## Exercise Set 2.1

Apply the Fundamental Property for Fractions. Multiply the numerator and denominator of each fraction by the given value for $C$ to find an equivalent fraction.

1. $\frac{3}{8}$, $C = 14$
2. $\frac{5}{7}$, $C = 3$
3. $\frac{2}{11}$, $C = 5$
4. $\frac{17}{19}$, $C = 6$
5. $\frac{41}{43}$, $C = 20$
6. $\frac{3}{74}$, $C = 35$
7. $\frac{38}{40}$, $C = 4$
8. $\frac{9}{22}$, $C = 15$
9. $\frac{71}{81}$, $C = 11$
10. $\frac{89}{90}$, $C = 5$

## 2.2 Prime Numbers

To change a fraction to its lowest terms, we divide its numerator and its denominator by the largest whole number which will divide both exactly. To find this number, we need to divide the numerator and the denominator into prime factors. **Factors** are numbers or expressions that are multiplied together to form a product. A **prime number** is a natural number whose only divisors are 1 and itself.

To find the prime numbers less than 100, make a chart like the one below. Make an "x" through every number that is not a prime. Since 1 is not prime by definition, put an "x" through 1. The smallest prime number is 2. Make an "x" through all multiples of 2, since any number that can be divided by 2 is not prime.

If we try dividing 3 by 2, we do not get a whole number as a quotient, so 3 is a prime number. Make an "x" through all multiples of 3, since any number that can be divided by 3 is not prime. Continue checking until each number has been shown to be either prime or composite. **Composite numbers** are natural numbers greater than 1 that are not prime.

The chart below has all numbers except prime numbers crossed out.

| | | | | | | | | | |
|---|---|---|---|---|---|---|---|---|---|
| X | 2 | 3 | X | 5 | X | 7 | X | X | X |
| 11 | X | 13 | X | X | X | 17 | X | 19 | X |
| X | X | 23 | X | X | X | X | X | 29 | X |
| 31 | X | X | X | X | X | 37 | X | X | X |
| 41 | X | 43 | X | X | X | 47 | X | X | X |
| X | X | 53 | X | X | X | X | X | 59 | X |
| 61 | X | X | X | X | X | 67 | X | X | X |
| 71 | X | 73 | X | X | X | X | X | 79 | X |
| X | X | 83 | X | X | X | X | X | 89 | X |
| X | X | X | X | X | X | 97 | X | X | X |

EXAMPLE 1: Which of the following numbers are prime numbers?

a) 14        d) 31

b) 17        e) 69

c) 75

SOLUTION:

a) 14 ÷ 2 = 7, so 14 is not prime.

b) 17 cannot be divided by any number other than 1 and 17, so 17 is prime.

c) 75 ÷ 5 = 15, so 75 is not prime. As 15 is not prime, 75 can be divided by any of the factors of 15, as well.

d) 31 cannot be divided by any number other than 1 and 31, so 31 is prime.

e) 69 ÷ 3 = 23, so 69 is not prime.

To write a number as a **prime factorization,** we write the number as a product of prime factors. We usually write the factors in a prime factorization in order from smallest to largest, from left to right. Every number has its own particular prime factorization, so no two numbers have the same prime factorization. A prime number's prime factorization is just the prime number itself.

EXAMPLE 2: Find the prime factorization for 42.

SOLUTION: 42 ÷ 2 = 21
21 ÷ 3 = 7

The prime factorization for 42 is: 2 (3) (7).

Numbers are relatively prime when there is no whole number other than 1 contained evenly in both of them. Thus, 8 and 15 are relatively prime because there is no number other than 1 that will divide both numbers without a remainder.

## Exercise Set 2.2

Write whether the following numbers are prime or composite.

**1.** 76      **4.** 11
**2.** 83      **5.** 39
**3.** 27

Write the prime factorization for the following numbers.

**6.** 12      **9.** 37
**7.** 63      **10.** 96
**8.** 85

## 2.3 Greatest Common Divisor

The largest number that is contained evenly in two or more other numbers is called the **greatest common divisor,** or GCD of these numbers.

An understanding of how to find the greatest common divisor of two or more numbers is necessary to perform operations with fractions.

To find the greatest common divisor of two or more numbers, find the prime factorization of each of the numbers. Multiply together the factors common to all of the numbers. The product of the common factors of all the numbers is the greatest common divisor.

EXAMPLE 1: Find the greatest common divisor of 42, 60, and 84.

SOLUTION: First, find the prime factorizations of 42, 60, and 84.

$$42 = 2\,(3)\,(7)$$
$$60 = 2\,(2)\,(3)\,(5)$$
$$84 = 2\,(2)\,(3)\,(7)$$

The factors common to all three numbers are 2 and 3, so the greatest common factor is 2 (3), or 6. No other factors will divide into all three numbers evenly.

Now that we know how to find the greatest common divisor, we can reduce fractions to their lowest terms. We put fractions in lowest terms by dividing the numerator and the denominator by their greatest common divisor.

EXAMPLE 2: Reduce $^{12}/_{30}$ to its lowest terms.

SOLUTION: Find the prime factorization for 12 and 30.

$$12 = 2\,(2)\,(3)$$
$$30 = 2\,(3)\,(5)$$

The factors common to both 12 and 30 are 2 and 3. The greatest common divisor is 2 (3), or 6. Divide both the numerator and the denominator by 6.

$$\frac{12}{30} = \frac{12 \div 6}{30 \div 6}$$

$$= \frac{2}{5}$$

EXAMPLE 3: Reduce $^{128}/_{288}$ to lowest terms.

SOLUTION: Find the prime factorization for 128 and 288.

$$128 = 2\,(2)\,(2)\,(2)\,(2)\,(2)$$

$$288 = 2\,(2)\,(2)\,(2)\,(3)\,(3)$$

The greatest common divisor is 2 (2) (2) (2) (2), or 32. Note that since the prime factorization shows us that we have repeated factors of 2 common to both 128 and 288, these repeated factors are used to find the greatest common divisor. Divide both the numerator and the denominator by 32.

$$\frac{128}{288} = \frac{128 \div 32}{288 \div 32}$$

$$= \frac{4}{9}$$

EXAMPLE 4: What is the greatest common divisor of 323 and 391?

SOLUTION: Find the prime factorizations of 323 and 391.

$$323 = 17\,(19)$$

$$391 = 17\,(23)$$

The only factor 323 and 391 have in common is 17, so 17 is their greatest common divisor.

To raise the denominator of a given fraction to a required denominator, divide the denominator of the given fraction into the required denominator, and then multiply both terms of the given fraction by the quotient.

EXAMPLE 5: Change $\frac{1}{4}$ to sixty-fourths.

SOLUTION: Since 64 is the required denominator, and 4 is the given denominator, divide 64 by 4 to get the quotient we use as a multiplier.

$$64 \div 4 = 16$$

Multiply both the 1 and the 4 by 16 to get the fraction with the required denominator.

$$\frac{1}{4} = \frac{1 \times 16}{4 \times 16}$$

$$= \frac{16}{64}$$

### Exercise Set 2.3

Find the greatest common divisor of the following groups of numbers.

| | |
|---|---|
| **1.** 12, 16, 28 | **9.** 33, 165 |
| **2.** 12, 72, 96 | **10.** 256, 608 |
| **3.** 14, 21, 35 | **11.** 24, 32, 104 |
| **4.** 15, 45, 81 | **12.** 36, 90, 153 |
| **5.** 32, 48, 80 | **13.** 48, 120, 168 |
| **6.** 48, 60 | **14.** 64, 256, 400 |
| **7.** 63, 99 | **15.** 81, 117, 120 |
| **8.** 54, 234 | |

Reduce each of the following fractions to lowest terms.

**16.** $^8/_{12}$ =

**17.** $^8/_{20}$ =

**18.** $^6/_{15}$ =

**19.** $^9/_{15}$ =

**20.** $^{12}/_{32}$ =

**21.** $^{16}/_{44}$ =

**22.** $^{10}/_{12}$ =

**23.** $^{13}/_{52}$ =

**24.** $^{10}/_{16}$ =

**25.** $^{18}/_{56}$ =

**26.** $^{20}/_{36}$ =

**27.** $^{42}/_{126}$ =

**28.** $^{15}/_{18}$ =

**29.** $^{144}/_{244}$ =

**30.** $^8/_{56}$ =

Change the following fractions to equivalent fractions using the indicated denominator.

**31.** $^1/_4$ to 8ths =

**32.** $^1/_3$ to 12ths =

**33.** $^2/_5$ to 20ths =

**34.** $^4/_9$ to 81sts =

**35.** $^3/_6$ to 48ths =

**36.** $^2/_7$ to 49ths =

**37.** $^4/_{32}$ to 64ths =

**38.** $^5/_{13}$ to 78ths =

**39.** $^3/_8$ to 24ths =

**40.** $^3/_5$ to 45ths =

**41.** $^2/_9$ to 36ths =

**42.** $^{11}/_{15}$ to 60ths =

**43.** $^{14}/_{25}$ to 75ths =

**44.** $^9/_{11}$ to 88ths =

**45.** $^5/_{12}$ to 96ths =

**46.** $^7/_{17}$ to 68ths =

## 2.4 Addition and Subtraction of Fractions

Just as we cannot add or subtract a number of feet from a number of yards until we convert both measurements to a common unit, we cannot add or subtract fractions until we convert them so they have a common denominator.

To add unlike fractions, first change them to equivalent fractions with the same denominator. Then add the numerators and put the sum over the common denominator.

EXAMPLE 1: Find $^1/_2$ plus $^3/_4$.

SOLUTION:

$$\frac{1}{2} + \frac{3}{4} = \frac{2}{4} + \frac{3}{4}$$

$$= \frac{2+3}{4}$$

$$= \frac{5}{4}, \text{ or } 1\frac{1}{4}$$

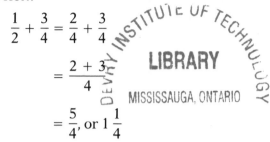 

EXAMPLE 2: Find $^1/_3$ plus $^3/_4$.

SOLUTION: The smallest number that is a multiple of both 3 and 4 is 12. We find that number by thinking of multiples of 4, the higher of the two denominators, as 4, 8, 12. We stop as soon as we come to the first number that is also a multiple of 3.

$$12 \div 3 = 4; \text{ so } \frac{1}{3} = \frac{4 \times 1}{4 \times 3} = \frac{4}{12}$$

$$12 \div 4 = 3; \text{ so } \frac{3}{4} = \frac{3 \times 3}{3 \times 4} = \frac{9}{12}$$

$$\frac{1}{3} + \frac{3}{4} = \frac{4}{12} + \frac{9}{12}$$

$$= \frac{4+9}{12}$$

$$= \frac{13}{12}, \text{ or } 1\frac{1}{12}$$

To change an improper fraction to a whole or mixed number, divide the numerator by the denominator and place the remainder over the denominator.

EXAMPLE 3: Convert $^{19}/_5$ to a mixed number.

SOLUTION: Divide 19 by 5.

$$\begin{array}{r} 3 \\ 5\overline{)19} \\ \underline{15} \\ 4 \end{array}$$

When we divide 19 by 5, we find that 5 divides into 19 three times with 4 left over. We can write this quotient as 3 with a remainder of 4, or as the mixed number $3\frac{4}{5}$.

To change a mixed number to an improper fraction, multiply the whole number by the denominator of the fraction, add the numerator to this product, and place the sum over the denominator.

EXAMPLE 4: Convert $2\frac{7}{8}$ to an improper fraction.

SOLUTION:

$$2\frac{7}{8} = \frac{(8 \times 2) + 7}{8}$$

$$= \frac{16 + 7}{8}$$

$$= \frac{23}{8}$$

EXAMPLE 5: Convert $4\frac{3}{5}$ to an improper fraction.

SOLUTION:

$$4\frac{3}{5} = \frac{(5 \times 4) + 3}{5}$$

$$= \frac{20 + 3}{5}$$

$$= \frac{23}{5}$$

To subtract unlike fractions, convert them to equivalent fractions with a common denominator. Then subtract the numerators and put the difference over the common denominator.

EXAMPLE 6: Subtract $5\frac{2}{3}$ from $3\frac{1}{6}$.

SOLUTION: First change $5\frac{2}{3}$ and $3\frac{1}{6}$ to improper fractions.

$$5\frac{2}{3} = \frac{(5 \times 3) + 2}{3} = \frac{17}{3}$$

$$3\frac{1}{6} = \frac{(3 \times 6) + 1}{6} = \frac{19}{6}$$

Now convert the fractions so that they have a common denominator of 6.

$$\frac{17}{3} = \frac{34}{6}$$

$$\frac{19}{6}$$

$$\frac{34}{6} - \frac{19}{6} = \frac{15}{6}$$

$$\frac{15}{6} = \frac{(2 \times 6) + 3}{6}$$

$$= 2\frac{1}{2}$$

We can convert the answer $\frac{15}{6}$ into the mixed number $2\frac{1}{2}$. The improper fraction $\frac{15}{6}$ and the mixed number $2\frac{1}{2}$ are equal.

## Exercise Set 2.4

Convert each fraction to a whole or mixed number.

1. $\frac{12}{5} =$
2. $\frac{14}{7} =$
3. $\frac{19}{12} =$
4. $\frac{43}{5} =$
5. $\frac{52}{8} =$
6. $\frac{114}{76} =$
7. $\frac{32}{14} =$
8. $\frac{28}{6} =$
9. $\frac{19}{4} =$
10. $\frac{21}{7} =$
11. $\frac{82}{41} =$
12. $\frac{96}{6} =$

Convert each mixed number to an improper fraction.

**13.** $2\frac{3}{4} =$     **19.** $19\frac{6}{7} =$

**14.** $3\frac{1}{4} =$     **20.** $16\frac{1}{6} =$

**15.** $4\frac{4}{5} =$     **21.** $12\frac{2}{7} =$

**16.** $5\frac{3}{5} =$     **22.** $13\frac{3}{7} =$

**17.** $12\frac{2}{3} =$     **23.** $14\frac{1}{5} =$

**18.** $18\frac{3}{4} =$     **24.** $22\frac{2}{5} =$

Add or subtract as indicated, and then write as a mixed number.

**25.** $\frac{3}{7} + \frac{2}{7} =$     **31.** $\frac{55}{11} - \frac{3}{22} =$

**26.** $\frac{1}{2} + \frac{4}{6} =$     **32.** $\frac{5}{16} - \frac{1}{8} =$

**27.** $3\frac{1}{5} - \frac{2}{15} =$     **33.** $\frac{11}{14} - \frac{2}{7} =$

**28.** $\frac{15}{45} + \frac{2}{5} =$     **34.** $\frac{1}{3} + \frac{2}{9} =$

**29.** $4\frac{2}{8} + 3\frac{1}{4} =$     **35.** $\frac{8}{13} - \frac{2}{26} =$

**30.** $\frac{31}{8} - \frac{12}{4} =$

## 2.5 Lowest Common Denominator

The term **lowest common denominator (LCD)** is used in connection with fractions. Numerically it is identical with the **least common multiple (LCM)** of the given denominators. The latter term has a more general use in mathematics. The least common multiple of two or more numbers is the smallest number that can be exactly divided by all the given numbers. Thus 12 is the least common multiple of 3 and 4; 45 is the least common multiple of 9 and 15. When applied to fractions, 12 is the least common denominator of $\frac{1}{3}$ and $\frac{1}{4}$; 45 is the least common denominator of $\frac{1}{9}$ and $\frac{1}{15}$.

To find the least common multiple of two numbers, first determine their greatest common divisor by finding the prime factorization; divide the numbers by this number, and then multiply together the resulting quotients and the greatest common divisor.

EXAMPLE 1: Find the lowest common denominator of $\frac{2}{9}$ and $\frac{6}{45}$.

SOLUTION: First, find the prime factorization of the denominators of both fractions. Then find the greatest common divisor of the two fractions.

$$9 = 3\,(3)$$
$$45 = 3\,(3)\,(5)$$

The greatest common divisor is 3 (3), or 9. Divide both 9 and 45 by 9.

$$9 \div 9 = 1$$
$$45 \div 9 = 5$$

Multiply the quotients 1 and 5 together, and then multiply the result by the greatest common divisor, 9.

$$1\,(5)\,(9) = 45$$

The lowest common denominator of $\frac{2}{9}$ and $\frac{6}{45}$ is 45. This means that if we wish to add or subtract the two fractions, the smallest possible denominator we can use is 45.

The lowest common denominator of more than two fractions can be found by factoring, but we must be careful not only to use the divisors that are contained in all the given numbers, but also any divisors that may be contained in two or more of them.

EXAMPLE 2: Use the lowest common denominator to find the sum of $\frac{3}{15}$, $\frac{1}{3}$, and $\frac{2}{45}$.

SOLUTION: Find the greatest common divisor of the denominators of the three fractions.

$$15 = 3\,(5)$$

$$3 = 3$$

$$45 = 3\,(3)\,(5)$$

Note that the prime factorization of 45 has all of the factors of 15, and of 3. Therefore, 45 is the smallest possible denominator that could be used to add the three fractions together. Rewrite the three fractions and then find the sum.

$$\frac{3}{15} = \frac{3 \times 3}{15 \times 3} = \frac{9}{45}$$

$$\frac{1}{3} = \frac{1 \times 15}{3 \times 15} = \frac{15}{45}$$

$$\frac{2}{45} = \frac{2}{45}$$

$$\frac{9}{45} + \frac{15}{45} + \frac{2}{45} = \frac{26}{45}$$

To find the least common multiple of more than two numbers by factoring, look for prime divisors that are common to two or more of the numbers. Divide the numbers by their common divisors. Then multiply together the common divisors and the final quotient.

To find the least common multiple of more than two numbers which cannot be readily factored, find the least common multiple of two of them, then the least common multiple of this result and another of the given numbers; continue in this way until all the original numbers have been used.

To subtract unlike fractions, first change them to equivalent fractions with their LCD. Then find the difference of the new numerators.

EXAMPLE 3: Find $\frac{3}{5} - \frac{1}{3}$.

SOLUTION: The LCD is 15.

$$\frac{3}{5} = \frac{9}{15}$$

$$\frac{1}{3} = \frac{5}{15}$$

$$\frac{3}{5} - \frac{1}{3} = \frac{9}{15} - \frac{5}{15}$$

$$= \frac{9 - 5}{15}$$

$$= \frac{4}{15}$$

To find the LCD when no two of the given denominators can be divided by the same number, multiply the denominators by each other; the result is the LCD.

EXAMPLE 4: Find $\frac{1}{2} + \frac{1}{3} + \frac{1}{5} + \frac{1}{7}$.

SOLUTION: $2 \times 3 \times 5 \times 7 = 210$

The LCD is 210.

$$\frac{1}{2} = \frac{105}{210}$$

$$\frac{1}{3} = \frac{70}{210}$$

$$\frac{1}{5} = \frac{42}{210}$$

$$\frac{1}{7} = \frac{30}{210}$$

$$\frac{105 + 70 + 42 + 30}{210} = \frac{247}{210}$$

$$= 1\frac{37}{210}$$

To add and subtract mixed numbers, treat the fractions separately; then add or subtract the results to or from the whole numbers.

EXAMPLE 5: Subtract $6\frac{3}{4}$ from $8\frac{1}{3}$.

SOLUTION: The LCD of $\frac{1}{3}$ and $\frac{3}{4}$ is 12.

$$\frac{1}{3} = \frac{4}{12}$$

$$\frac{3}{4} = \frac{9}{12}$$

$$8\frac{4}{12} = 7 + \frac{12}{12} + \frac{4}{12}$$

$$= 7\frac{16}{12}$$

$$7\frac{16}{12}$$

$$-6\frac{9}{12}$$

$$1\frac{7}{12}$$

Since $\frac{9}{12}$ is greater than $\frac{4}{12}$, it is necessary to borrow 1 from the 8, which then becomes 7; so $7 + \frac{12}{12} + \frac{4}{12} = 7\frac{16}{12}$. When we subtract $6\frac{9}{12}$, the result is $1\frac{7}{12}$.

## Exercise Set 2.5

Use the lowest common denominator method to find the sum or difference.

1. $\frac{7}{8} + \frac{3}{4}$
2. $\frac{8}{9} - \frac{2}{3}$
3. $\frac{7}{8} - \frac{3}{5}$
4. $\frac{5}{6} + \frac{8}{9}$
5. $\frac{3}{4} + \frac{5}{12} - \frac{2}{3}$
6. $\frac{7}{8} - \frac{1}{2} - \frac{1}{4}$
7. $5\frac{1}{2} + 3\frac{3}{4}$
8. $3\frac{2}{3} + 1\frac{5}{6} + 2\frac{1}{4}$
9. $15\frac{1}{9} + 8\frac{5}{6}$
10. $12\frac{7}{8} - 6\frac{1}{3}$
11. $9\frac{1}{3} - 7\frac{3}{4}$
12. $16\frac{3}{8} - 9\frac{5}{6}$

## 2.6 Multiplication and Division of Fractions

To multiply a fraction by a whole number, multiply the numerator by the whole number. The product will be the new numerator over the old denominator.

EXAMPLE 1: Multiply 6 by $\frac{2}{3}$.

SOLUTION:

$$6 \times \frac{2}{3} = \frac{6 \times 2}{3}$$

$$= \frac{12}{3}$$

$$= 4$$

To divide a fraction by a whole number, multiply the denominator of the fraction by the whole number. The quotient will be the old numerator over the new denominator.

EXAMPLE 2: Divide $\frac{2}{3}$ by 5.

SOLUTION:

$$\frac{2}{3} \div 5 = \frac{2}{3 \times 5}$$

$$= \frac{2}{15}$$

EXAMPLE 3: Divide $\frac{1}{5}$ by 2.

SOLUTION:

$$\frac{1}{5} \div 2 = \frac{1}{5 \times 2} = \frac{1}{10}$$

To multiply one fraction by other fractions, place the product of the numerators over the product of the denominators; then reduce.

EXAMPLE 4: Find the product of $\frac{2}{3}$, $\frac{1}{5}$, and $\frac{3}{6}$.

SOLUTION:

$$\frac{2}{3} \times \frac{1}{5} \times \frac{3}{6} = \frac{2 \times 1 \times 3}{3 \times 5 \times 6}$$

$$= \frac{6}{90}$$

$$= \frac{1}{15}$$

To multiply mixed numbers, change the mixed numbers to improper fractions.

EXAMPLE 5: Find the product of $1\frac{3}{4}$, $2\frac{2}{3}$, and $1\frac{1}{2}$.

SOLUTION:

$$1\frac{3}{4} \times 2\frac{2}{3} \times 1\frac{1}{2} = \frac{7}{4} \times \frac{8}{3} \times \frac{3}{2}$$

$$= \frac{7 \times 8 \times 3}{4 \times 3 \times 2}$$

$$= \frac{168}{24}$$

$$= 7$$

To divide a whole number or a fraction by a fraction, invert the divisor and multiply. An inverted fraction is the **reciprocal** of the original fraction. For instance, $\frac{2}{3}$ is the reciprocal of $\frac{3}{2}$, and $\frac{2}{5}$ is the reciprocal of $\frac{5}{2}$.

The reciprocal of a whole number is found by putting a 1 over the whole number. Thus, the reciprocal of 2 is $\frac{1}{2}$, and the reciprocal of 16 is $\frac{1}{16}$.

EXAMPLE 6: Divide $\frac{1}{2}$ by $\frac{2}{3}$.

SOLUTION: Multiply $\frac{1}{2}$ by the reciprocal of $\frac{2}{3}$.

$$\frac{1}{2} \div \frac{2}{3} = \frac{1}{2} \times \frac{3}{2}$$

$$= \frac{3}{4}$$

EXAMPLE 7: Divide $\frac{4}{7}$ by $\frac{2}{5}$.

SOLUTION: Multiply $\frac{4}{7}$ by the reciprocal of $\frac{2}{5}$.

$$\frac{4}{7} \div \frac{2}{5} = \frac{4}{7} \times \frac{5}{2}$$

$$= \frac{20}{14}$$

$$= 1\frac{6}{14}$$

$$= 1\frac{3}{7}$$

**PRACTICALLY SPEAKING 2.6**

Alan is making spice cookies for a party. He wishes to triple the recipe, as he wants to make sure there are enough cookies for everyone.

Here is the recipe Alan must triple:

**Spice Cookies**

2 cups flour

1 egg

$\frac{1}{2}$ cup honey

$\frac{1}{2}$ teaspoon grated lemon peel

$\frac{1}{4}$ teaspoon cinnamon

$\frac{1}{4}$ teaspoon ginger

$\frac{1}{2}$ teaspoon allspice

$\frac{1}{3}$ teaspoon nutmeg

$\frac{1}{3}$ teaspoon cloves

$\frac{1}{4}$ teaspoon baking soda

1. What does Alan need to do in order to triple the amount of cookies for his party?

2. How much cinnamon does he need in the tripled recipe?

3. How much nutmeg does he need in the tripled recipe?

4. How much honey does he need in the tripled recipe?

See Appendix F for the answers.

# Exercise Set 2.6

**1.** $\frac{3}{7} \times \frac{3}{5} =$

**2.** $\frac{3}{8} \times \frac{2}{3} =$

**3.** $\frac{2}{15} \times \frac{4}{91} \times \frac{7}{8} \times 5 =$

**4.** $\frac{14}{15} \times \frac{9}{28} =$

**5.** $\frac{3}{8} \times 12 =$

**6.** $\frac{15}{16} \times \frac{7}{90} =$

**7.** $\frac{5}{6} \times \frac{5}{9} =$

**8.** $\frac{7}{27} \times \frac{9}{14} =$

**9.** $18 \div \frac{1}{2} =$

**10.** $\frac{3}{5} \div \frac{1}{15} =$

**11.** $\frac{2}{3} \div \frac{1}{2} =$

**12.** $\frac{4}{7} \div \frac{1}{8} =$

**13.** $1\frac{2}{3} \div \frac{3}{4} =$

**14.** $2\frac{1}{2} \div 1\frac{3}{4} =$

**15.** $3\frac{1}{2} \div \frac{1}{4} =$

**16.** $1\frac{1}{8} \div \frac{3}{16} =$

## 2.7 Simplification of Fractions

Simplifying fractions, or reducing fractions, is a short cut in the process of multiplication of fractions. If we write both the numerator and denominator as the product of prime factors, then we can simplify the multiplication by dividing out common factors.

Thus in the calculation

$$\frac{2}{4} \times \frac{3}{6} \times \frac{4}{3} =$$

$$\frac{\cancel{2}}{2 \times \cancel{2}} \times \frac{\cancel{3}}{\cancel{2} \times 3} \times \frac{\cancel{2} \times 2}{\cancel{3}}$$

$$= \frac{1}{3}$$

the 4's can be divided out as the prime factors 2 × 2. The 3's can be divided out, and the 2 in the numerator is contained 3 times in the 6 in the denominator, leaving 3 when the prime factor 2 is divided out.

Simplification of fractions can be applied only to multiplication and division of fractions: never to addition and subtraction of fractions.

EXAMPLE 1: Find the product of $^{10}/_{25}$, $^{4}/_{3}$, and $^{12}/_{8}$, first reducing fractions.

SOLUTION:

$$\frac{10}{25} \times \frac{4}{3} \times \frac{12}{8} = \frac{2 \times \cancel{5}}{5 \times \cancel{5}} \times \frac{2 \times \cancel{2}}{\cancel{3}}$$

$$\times \frac{\cancel{2} \times \cancel{2} \times \cancel{3}}{\cancel{2} \times \cancel{2} \times \cancel{2}}$$

$$= \frac{4}{5}$$

Reducing factors can save us time when multiplying or dividing. If all factors common to both the numerator and the denominator are divided out, the result will always be in lowest terms.

A complex fraction is a fraction whose numerator, denominator, or both, contains a fraction. For instance,

$$\frac{2 - \dfrac{4}{16}}{4 + \dfrac{3}{5}}$$

is a complex fraction.

To simplify complex fractions, convert the numerators and denominators to simple fractions. Then follow the rules for adding, subtracting, multiplying, or dividing simple fractions.

EXAMPLE 2: Simplify the complex fraction

$$\frac{\dfrac{1}{2} + \dfrac{3}{4}}{2 + \dfrac{1}{2}}$$

SOLUTION:

$$\frac{\dfrac{1}{2} + \dfrac{3}{4}}{2 + \dfrac{1}{2}} = \frac{\dfrac{5}{4}}{\dfrac{5}{2}}$$

$$= \frac{\overset{1}{\cancel{5}}}{\underset{2}{\cancel{4}}} \times \frac{\overset{1}{\cancel{2}}}{\underset{1}{\cancel{5}}}$$

$$= \frac{1}{2}$$

EXAMPLE 3: Simplify the complex fraction

$$\frac{1\frac{2}{3} \times 2\frac{1}{4}}{\frac{1}{2} \div \frac{2}{3}}$$

SOLUTION:

$$\frac{1\frac{2}{3} \times 2\frac{1}{4}}{\frac{1}{2} \div \frac{2}{3}} = \frac{\frac{5}{3} \times \frac{9}{4}}{\frac{1}{2} \times \frac{3}{2}}$$

$$= \frac{\frac{45}{12}}{\frac{3}{4}}$$

$$= \frac{\overset{5}{\cancel{45}}}{\underset{3}{\cancel{12}}} \times \frac{\overset{1}{\cancel{4}}}{\underset{1}{\cancel{3}}}$$

$$= 5$$

To multiply a whole number and a mixed number together, perform separate multiplications and add the results.

EXAMPLE 4: Multiply 17 by 6¾.

SOLUTION: We multiply 17 first by 6, the whole number part of the multiplier, and then by the fractional part, ¾. Then we add the results together.

$$\begin{array}{r} 17 \\ 6\frac{3}{4} \\ \hline 102 \\ 12\frac{3}{4} \\ \hline 114\frac{3}{4} \end{array}$$

EXAMPLE 5: Multiply 17⅗ by 4.

SOLUTION: We first multiply the ⅗ in the multiplicand by 4, the multiplier, and the result is ¹²⁄₅, or 2⅖. We then multiply the whole number part, 17, by 4. Finally, we add the two products.

$$\begin{array}{r} 17\frac{3}{5} \\ 4 \\ \hline 2\frac{2}{5} \\ 68 \\ \hline 70\frac{2}{5} \end{array}$$

To divide a mixed number by a whole number, or a whole number by a mixed number, first convert the mixed number to an improper fraction. Then divide as usual.

EXAMPLE 6: Divide 17⅜ by 6.

SOLUTION: Convert 17⅜ to an improper fraction.

$$17\frac{3}{8} = \frac{(17 \times 8) + 3}{8}$$

$$= \frac{139}{8}$$

Divide ¹³⁹⁄₈ by 6.

$$\frac{139}{8} \div 6 = \frac{139}{8} \times \frac{1}{6}$$

$$= \frac{139}{48}$$

$$= 2\frac{43}{48}$$

EXAMPLE 7: Divide 25 by $4\frac{3}{5}$.

SOLUTION: First convert $4\frac{3}{5}$ to an improper fraction.

$$4\frac{3}{5} = \frac{(4 \times 5) + 3}{5}$$

$$= \frac{23}{5}$$

Divide 25 by $\frac{23}{5}$.

$$25 \div \frac{23}{5} = 25 \times \frac{5}{23}$$

$$= \frac{125}{23}$$

$$= 5\frac{10}{23}$$

## Exercise Set 2.7

1. $\dfrac{\frac{2}{3} + \frac{1}{4} + \frac{1}{2}}{\frac{5}{8} - \frac{1}{6} - \frac{1}{4}} =$

2. $\dfrac{\frac{2}{3} \div \frac{1}{5}}{\frac{1}{4} \times \frac{1}{3}} =$

3. $\dfrac{2\frac{1}{2} + 3\frac{1}{3}}{4 + \frac{2}{3}} =$

4. $\dfrac{\frac{1}{4} \text{ of } 8}{1\frac{1}{2} \text{ of } 3} =$

5. $\dfrac{\frac{1}{5} + \frac{8}{14} + \frac{3}{7}}{\frac{3}{7} \div 10} =$

6. $\dfrac{\frac{1}{6} + \frac{2}{5} + \frac{3}{4} + \frac{1}{3}}{\frac{7}{12} - \frac{1}{6} + \frac{1}{3}} =$

7. $9\frac{3}{8} \times 5 =$

8. $12\frac{3}{5} \times 7 =$

9. $9 \times 8\frac{11}{12} =$

10. $10 \times 7\frac{1}{9} =$

11. $11\frac{6}{7} \times 8 =$

12. $7\frac{6}{11} \times 5 =$

13. $23\frac{7}{12} \times 6 =$

14. $8\frac{3}{8} \times 5 =$

15. $9 \times 6\frac{3}{8} =$

16. $12 \times 637\frac{1}{2} =$

17. $17\frac{3}{5} \div 7 =$

18. $18\frac{3}{7} \div 8 =$

19. $27\frac{11}{12} \div 9 =$

20. $31\frac{1}{10} \div 11 =$

21. $78\frac{4}{5} \div 12 =$

22. $36 \div 9\frac{7}{8} =$

23. $97 \div 13\frac{11}{12} =$

24. $342 \div 14\frac{47}{131} =$

25. $113 \div 21\frac{1}{7} =$

26. $19 \div 2\frac{3}{7} =$

## Chapter 2 Glossary

**Complex Fractions** A fraction where either the numerator or the denominator, or both, are fractions or mixed numbers.

**Composite Number** A natural number greater than 1 that is not prime.

**Denominator** The divisor in a fraction.

**Factors** Numbers or expressions that are multiplied together to form a product.

**Fraction** The quotient of two whole numbers.

**Greatest Common Divisor** The largest number contained in two or more other numbers.

**Improper Fraction** A fraction with a value greater than 1.

**Lowest Common Denominator** The smallest denominator possible for two or more denominators.

**Lowest Terms** When a fraction is written with the smallest possible numbers for both its numerator and denominator.

**Mixed Number** A whole number and a fraction written together with the understanding that they are to be added together.

**Numerator** The dividend in a fraction.

**Prime Factorization** A number written as a product of prime factors.

**Prime Number** A natural number whose only divisors are 1 and itself.

**Proper Fraction** A fraction with a value less than 1.

**Reciprocal** The multiplicative inverse of a fraction or whole number.

**Relatively Prime Numbers** When there is no whole number other than 1 that divides the numbers without leaving a remainder.

**Simple Fraction** A fraction where both the numerator and the denominator are whole numbers.

## Chapter 2 Test

For each problem, five answers are given. Only one answer is correct. After you solve each problem, check the answer that agrees with your solution.

1. Choose the prime number.

   A) 34   D) 21
   B) 13   E) 33
   C) 69

2. Choose the greatest common divisor, or GCD, of 96, 39, and 42.

   A) 6   D) 3
   B) 2   E) 4
   C) 7

3. Reduce $^{16}/_{52}$ to lowest terms.

   A) $^4/_{13}$   D) $^4/_{25}$
   B) $^2/_{12}$   E) $^1/_3$
   C) $^8/_{26}$

4. Find a fraction equivalent to $^3/_8$.

   A) $^9/_{32}$   D) $^{13}/_{24}$
   B) $^7/_{11}$   E) $^{16}/_{24}$
   C) $^9/_{24}$

5. Find the sum of $1^1/_2$ and $3^5/_8$.

   A) $4^6/_8$   D) $4^3/_8$
   B) $4^1/_8$   E) $5^1/_8$
   C) $5^6/_8$

6. Find the difference between $^{17}/_8$ and $^3/_4$.

   A) $^{14}/_8$   D) $^{14}/_4$
   B) $^{11}/_8$   E) $^{11}/_4$
   C) $^{20}/_8$

7. Find the lowest common denominator of the fractions $^5/_{32}$, $^7/_8$, and $^3/_{16}$.

   A) 8   D) 4
   B) 64   E) 16
   C) 32

8. Write $5^3/_8$ as an improper fraction.

   A) $^8/_8$   D) $^{15}/_8$
   B) $^{43}/_8$   E) $^{37}/_8$
   C) $^{83}/_8$

9. Find the difference between $^{52}/_{56}$ and $^3/_7$.

   A) $^{28}/_{56}$   D) $^{37}/_{56}$
   B) $^{22}/_{56}$   E) $^{49}/_{56}$
   C) $^{40}/_{56}$

**10.** Find the prime factorization of 78.

A) $2 \times 3 \times 3$    D) $3 \times 7 \times 13$
B) $2 \times 5 \times 7$    E) $2 \times 7 \times 11$
C) $2 \times 3 \times 13$

**11.** Find the product of $\frac{3}{8}$ and $\frac{3}{5}$.

A) $\frac{6}{13}$    D) $\frac{1}{4}$
B) $\frac{11}{40}$    E) $\frac{6}{40}$
C) $\frac{9}{40}$

**12.** Find the product of $\frac{1}{4}$ and $\frac{13}{17}$.

A) $\frac{13}{21}$    D) $\frac{12}{21}$
B) $\frac{13}{68}$    E) $\frac{24}{34}$
C) $\frac{52}{68}$

**13.** Divide $\frac{3}{4}$ by $\frac{1}{8}$, and reduce to lowest terms if possible.

A) 6    D) 2
B) 3    E) $\frac{3}{2}$
C) $\frac{3}{32}$

**14.** Divide $\frac{1}{21}$ by $\frac{3}{29}$ and reduce to lowest terms if possible.

A) $\frac{29}{21}$    D) $\frac{3}{609}$
B) $\frac{29}{63}$    E) $\frac{7}{29}$
C) $\frac{7}{63}$

**15.** Reduce $\frac{6}{132}$ to lowest terms.

A) $\frac{3}{32}$    D) $\frac{6}{132}$
B) $\frac{1}{51}$    E) $\frac{1}{22}$
C) $\frac{1}{15}$

**16.** How many sheets of metal $\frac{1}{32}$ inches thick are there in a pile $25\frac{1}{2}$ inches high?

A) 550    D) 816
B) 105    E) 275
C) 408

**17.** If $\frac{1}{3}$ the length of a beam is 10 feet, how long is the entire beam?

A) 3 feet    D) 9 feet
B) 30 feet    E) 27 feet
C) $3\frac{1}{3}$ feet

*Note:* To find the value of the whole when the fractional part is given, invert the fraction and multiply it by the given part.

**18.** If it takes 5 hours to wax $\frac{2}{3}$ of a car, how long will it take to wax the whole car?

A) $7\frac{1}{2}$ hours    D) 10 hours
B) $3\frac{1}{2}$ hours    E) $\frac{27}{3}$ hours
C) 7 hours

**19.** An aviator made 3 flights. The first was 432 miles, the second was only $\frac{1}{2}$ that distance and the third $\frac{1}{3}$ of the original distance. How far would he have to go on a fourth flight to equal $\frac{1}{2}$ the distance covered by the second and third trips?

A) 180 miles    D) 275 miles
B) 450 miles    E) 235 miles
C) 360 miles

**20.** The distance between New York and California is 3,000 miles. Two trains leave the two cities at the same time. One train travels at the rate of $62\frac{3}{5}$ miles an hour, the other at $69\frac{4}{5}$ miles per hour. How far apart will the two trains be at the end of 5 hours?

A) 662 miles    D) 1,842 miles
B) 2,338 miles    E) 3,560 miles
C) 2,220 miles

# Decimals

## 3.1 Decimals

**Decimal fractions** are a special way of writing proper fractions that have denominators beginning with 1 and ending with one or more zeros. Thus, when written as decimal fractions,

$$\frac{1}{10}, \frac{2}{100}, \frac{3}{1,000}, \frac{4}{10,000}, \text{ and } \frac{5}{100,000}$$

become

0.1, 0.02, 0.003, 0.0004, and 0.00005.

The period before the digits is the **decimal point;** the digits following it stand for certain decimal places.

The word decimal means relating to the number 10. To calculate fractions by decimals is simply to extend into the field of fractions the same method of counting that we employ when dealing with whole numbers.

Read the number after the decimal point as a whole number and give it the name of its last decimal place.

0.135 is read as one hundred thirty-five thousandths

4.18 is read as four and eighteen hundredths

Another way to read decimals is:

0.135 is point, one-three-five

4.18 may be read four, point, one-eight

| PLACE OF DIGIT | HOW TO READ IT | EXAMPLE |
|---|---|---|
| First decimal place | Tenths | $0.3$ is $\dfrac{3}{10}$ |
| Second decimal place | Hundredths | $0.03$ is $\dfrac{3}{100}$ |
| Third decimal place | Thousandths | $0.003$ is $\dfrac{3}{1,000}$ |
| Fourth decimal place | Ten thousandths | $0.0003$ is $\dfrac{3}{10,000}$ |
| Fifth decimal place | Hundred thousandths | $0.00003$ is $\dfrac{3}{100,000}$ |

## Exercise Set 3.1

Write out each decimal in words.

1. 0.265
2. 0.79
3. 0.842
4. 0.3911
5. 0.5017
6. 0.053
7. 0.00061
8. 0.2001
9. 0.45
10. 0.0001

Write the decimal numbers described below.

11. Two hundred and fifty thousandths
12. Four and twenty-three hundredths
13. Twelve and forty thousandths
14. Four thousand sixty-two millionths
15. Seven hundred fifteen and eight tenths
16. Three hundred thousandths
17. Nineteen thousand and thirty-seven ten thousandths
18. Eighty hundredths
19. Sixteen ten thousandths
20. Fifty-one thousandths

## 3.2 Converting Fractions to Decimals

We can convert some fractions to decimals by placing the decimal point and the correct number of zeros before the numerator, and eliminating the denominator. We can do this when the denominators are 10's or some multiple of 10, such as 100, 1,000, or 10,000.

EXAMPLE 1: Convert 23/100 to a decimal.

SOLUTION: 23/100 = 0.23

To change any common fraction into decimals, divide the numerator by the denominator and write the quotient in decimal form.

EXAMPLE 2: Convert 3/5 to a decimal.

SOLUTION:
$$5\overline{)3.0} \quad 0.6$$

The decimal equivalent to 3/5 is 0.6.

EXAMPLE 3: Convert 3/8 to a decimal.

SOLUTION:   $\begin{array}{r} 0.375 \\ 8\overline{)3.000} \end{array}$

The decimal equivalent to 3/8 is 0.375.

Here is a table showing the decimal equivalents for fractions expressed as sixty-fourths. Note that the fractions in the table are all written in lowest terms. All of these are frequently used in technical work.

**DECIMAL EQUIVALENTS OF SIXTY-FOURTHS**

| FRACTION | DECIMAL | FRACTION | DECIMAL |
|---|---|---|---|
| $\frac{1}{64}$ | .....0.015625 | $\frac{33}{64}$ | .....0.515625 |
| $\frac{1}{32}$ | ..........0.03125 | $\frac{17}{32}$ | ..........0.53125 |
| $\frac{3}{64}$ | .....0.046875 | $\frac{35}{64}$ | .....0.546875 |
| $\frac{1}{16}$ | ...................0.0625 | $\frac{9}{16}$ | ...................0.5625 |
| $\frac{5}{64}$ | .....0.078125 | $\frac{37}{64}$ | .....0.578125 |
| $\frac{3}{32}$ | ..........0.09375 | $\frac{19}{32}$ | ..........0.59375 |
| $\frac{7}{64}$ | .....0.109375 | $\frac{39}{64}$ | .....0.609375 |
| $\frac{1}{8}$ | ...........................0.125 | $\frac{5}{8}$ | ...........................0.625 |
| $\frac{9}{64}$ | .....0.140625 | $\frac{41}{64}$ | .....0.640625 |
| $\frac{5}{32}$ | ..........0.15625 | $\frac{21}{32}$ | ..........0.65625 |
| $\frac{11}{64}$ | .....0.171875 | $\frac{43}{64}$ | .....0.671875 |
| $\frac{3}{16}$ | ...................0.1875 | $\frac{11}{16}$ | ...................0.6875 |
| $\frac{13}{64}$ | .....0.203125 | $\frac{45}{64}$ | .....0.703125 |
| $\frac{7}{32}$ | ..........0.21875 | $\frac{23}{32}$ | ..........0.71875 |
| $\frac{15}{64}$ | .....0.234375 | $\frac{47}{64}$ | .....0.734375 |

**DECIMAL EQUIVALENTS OF SIXTY-FOURTHS (*Continued*)**

| FRACTION | DECIMAL | FRACTION | DECIMAL |
|---|---|---|---|
| $\frac{1}{4}$ | 0.25 | $\frac{3}{4}$ | 0.75 |
| $\frac{17}{64}$ | 0.265625 | $\frac{49}{64}$ | 0.765625 |
| $\frac{9}{32}$ | 0.28125 | $\frac{25}{32}$ | 0.78125 |
| $\frac{19}{64}$ | 0.296875 | $\frac{51}{64}$ | 0.796875 |
| $\frac{5}{16}$ | 0.3125 | $\frac{13}{16}$ | 0.8125 |
| $\frac{21}{64}$ | 0.328125 | $\frac{53}{64}$ | 0.828125 |
| $\frac{11}{32}$ | 0.34375 | $\frac{27}{32}$ | 0.84375 |
| $\frac{23}{64}$ | 0.359375 | $\frac{55}{64}$ | 0.859375 |
| $\frac{3}{8}$ | 0.375 | $\frac{7}{8}$ | 0.875 |
| $\frac{25}{64}$ | 0.390625 | $\frac{57}{64}$ | 0.890625 |
| $\frac{13}{32}$ | 0.40625 | $\frac{29}{32}$ | 0.90625 |
| $\frac{27}{64}$ | 0.421875 | $\frac{59}{64}$ | 0.921875 |
| $\frac{7}{16}$ | 0.4375 | $\frac{15}{16}$ | 0.9375 |
| $\frac{29}{64}$ | 0.453125 | $\frac{61}{64}$ | 0.953125 |
| $\frac{15}{32}$ | 0.46875 | $\frac{31}{32}$ | 0.96875 |
| $\frac{31}{64}$ | 0.484375 | $\frac{63}{64}$ | 0.984375 |
| $\frac{1}{2}$ | 0.5 | 1 | 1. |

## Exercise Set 3.2

Rewrite each fraction as a decimal.

1. $\frac{3}{10} =$

2. $\frac{5}{100} =$

3. $\frac{321}{1000} =$

4. $12\frac{1}{100} =$

5. $124\frac{3}{10000} =$

6. $18\frac{7}{10} =$

7. $\frac{300}{1000} =$

8. $\frac{145}{100} =$

9. $\frac{223}{10} =$

10. $\frac{4330}{1000} =$

Using the table, find the decimal equivalents to the nearest thousandth for the following fractions.

| | | | |
|---|---|---|---|
| **11.** | $\frac{1}{2}$ = | **16.** | $\frac{17}{32}$ = |
| **12.** | $\frac{3}{4}$ = | **17.** | $\frac{28}{32}$ = |
| **13.** | $\frac{3}{8}$ = | **18.** | $\frac{14}{16}$ = |
| **14.** | $\frac{5}{16}$ = | **19.** | $\frac{22}{32}$ = |
| **15.** | $\frac{9}{16}$ = | **20.** | $\frac{56}{64}$ = |

## 3.3 Converting Decimals to Fractions

To write any decimal as a fraction, write the number after the decimal point as a numerator, with a denominator beginning with one and having as many zeros as there are numbers after the decimal point in the original decimal. Then reduce the fraction, if possible. Remember that whole numbers stay the same.

EXAMPLE 1: Convert 0.425 to a fraction.

SOLUTION: Write 425 as the numerator of the fraction. Write a 1 followed by three zeros, or 1,000, as the denominator. The decimal 0.425 is equivalent to 425/1,000, or 17/40 in lowest terms.

EXAMPLE 2: Convert 54.62 to a mixed number.

SOLUTION: Write 62 as the numerator of the fraction. Write a 1 followed by two zeros, or 100, as the denominator. The decimal 0.62 is equivalent to 62/100, or 31/50. The decimal 54.62 equals 54 31/50.

## Exercise Set 3.3

Rewrite each number as a common fraction, or as a mixed number in lowest terms.

| | | | |
|---|---|---|---|
| **1.** | 0.01 = | **6.** | 0.0008 = |
| **2.** | 0.5 = | **7.** | 0.0608 = |
| **3.** | 0.625 = | **8.** | 0.2341 = |
| **4.** | 2.10 = | **9.** | 0.04329 = |
| **5.** | 23.450 = | **10.** | 18.0200 = |

## 3.4 Addition and Subtraction of Decimals

To add or subtract decimals, place the numbers in a column with the decimal points lined up. Add or subtract as for whole numbers. Line up the decimal point in the result under the decimal points in the column.

EXAMPLE 1: Find the sum of 2.43, 1.485, 0.3, 12.02, and 0.074.

SOLUTION: Since 1.485 and 0.074 are three-place numbers, we write zeros after 2.43, 0.3, and 12.02. This does not change the value but helps to avoid errors.

```
 2.43     or      2.430
 1.485            1.485
 0.3              0.300
12.02            12.020
 0.074            0.074
─────            ──────
16.309           16.309
```

EXAMPLE 2: Find the difference between 17.29 and 6.147.

SOLUTION: As above, we add a zero to 17.29 to make it a three-place number. This does not change the value, and is not strictly necessary but helps to avoid errors.

```
 17.29    or     17.290
- 6.147          - 6.147
──────           ───────
 11.143           11.143
```

## Exercise Set 3.4

Add or subtract the decimals as indicated.

1. $0.2 + 0.07 + 0.5 =$
2. $2.6 + 22.4 + 0.03 =$
3. $22.8 + 5.099 + 613.2 =$
4. $0.005 + 5 + 16.2 + 0.96 =$
5. $15.4 + 22 + 0.01 + 1.48 =$
6. $28.74 - 16.32 =$
7. $0.005 - 0.0005 =$
8. $1.431 - 0.562 =$
9. $1.0020 - 0.2 =$
10. $8.04 - 7.96 =$
11. $72.306 + 18.45 - 27.202 =$
12. $14 - 6.3 + 2.739 =$
13. $27.65 + 18.402 - 2.39 + 7.63 =$
14. $18.0006 + 14.005 + 12.34 =$
15. $93.8 - 16.4327 - 20.009 =$
16. $14.29 - 6.305 - 3.47265 =$

## 3.5 Multiplication of Decimals

To multiply decimals, proceed as in multiplication of whole numbers. But in the product, beginning at the right, count off as many decimal places as there are in the multiplier and in the multiplicand together. Then place the decimal point.

EXAMPLE 1: Multiply 3.12 by 0.42.

SOLUTION: Since there are a total of four decimal places when we add together those in the multiplier and in the multiplicand, we start at the right and count four places; hence we put the decimal point off to the left of the 3, which marks the fourth place counted off.

```
    3.12      (Multiplicand: two decimal places.)
 ×  0.42      (Multiplier: two decimal places.)
   624
  1248
 1.3104       (Product has two plus two, or four
                decimal places.)
```

EXAMPLE 2: Multiply 0.214 by 0.303.

SOLUTION: There are a total of six places in the multiplier and in the multiplicand, but there are only five numbers in the product; therefore we prefix a zero at the left end, and place our decimal point before it to give the required six decimal places. If we needed eight places and the answer came out to five places, we would prefix three zeros and place the decimal point to the left of them.

```
     0.214
 ×   0.303
     642
    6420
  .?64842  = 0.064842
```

To multiply a decimal by any multiple of 10, move the decimal point as many places to the right as there are zeros in the multiplier.

EXAMPLE 3: Multiply 0.31 by 100.

SOLUTION: Since 100 has two zeros, move the decimal point in 0.31 two places to the right.

$$0.31 \times 100 = 31$$

EXAMPLE 4: Multiply 0.0021 by 1,000.

SOLUTION: Since 1,000 has three zeros, move the decimal point in 0.0021 three places to the right. We drop the zero in front, since a zero in front of a whole number is meaningless.

$$0.0021 \times 1,000 = 2.1$$

To divide a decimal or a whole number by 10 or a multiple of 10 such as 100 or 10,000, we move the decimal point as many places to the left as there are zeros in the divisor.

EXAMPLE 5: Divide 42 by 10.

SOLUTION: Since 10 has one zero, we move the decimal point one place to the left.

$$42 \div 10 = 4.2$$

EXAMPLE 6: Divide 61 by 1,000.

SOLUTION: Since 1,000 has three zeros, we move the decimal point three places to the left. Insert a zero to the left of 61 to give the new decimal the required number of decimal places.

$$61 \div 1,000 = .061$$

## Exercise Set 3.5

Multiply or divide as indicated.

1. $18.5 \times 4 =$
2. $3.9 \times 2.4 =$
3. $45 \times .72 =$
4. $143 \times .214 =$
5. $.56 \times .74 =$
6. $.224 \times .302 =$
7. $7.43 \times .132 =$
8. $.021 \times .204 =$
9. $.601 \times .003 =$
10. $.014 \times .0064 =$
11. $13.2 \times 2.475 =$
12. $.132 \times 2.475 =$
13. $.236 \times 12.13 =$
14. $9.06 \times .045 =$
15. $.008 \times 751.1 =$
16. $8.7 \times 10 =$
17. $.0069 \times 10 =$
18. $95.6 \times 100 =$
19. $.0453 \times 100 =$

**20.** $4.069 \times 1,000 =$
**21.** $.000094 \times 10,000 =$
**22.** $9.2 \times 10 =$
**23.** $7.49 \times 100 =$
**24.** $534.79 \div 100 =$
**25.** $492.568 \div 1,000 =$
**26.** $24.9653 \div 1,000 =$
**27.** $5.908 \div 100 =$
**28.** $.07156 \div 1,000 =$
**29.** $4956.74 \div 10,000 =$
**30.** $.038649 \div 100,000 =$

## 3.6 Division of Decimals

In division, a quotient is not changed when the dividend and divisor are both multiplied by the same number.

EXAMPLE 1: Divide 7.2 by 0.9.

SOLUTION: If we multiply both the dividend and the divisor by 10, the new division lets us divide by whole numbers.

$$7.2 \times 10 = 72$$
$$0.9 \times 10 = 9$$
$$72 \div 9 = 8$$

Check: $8 \times 0.9 = 7.2$

To divide a decimal by a whole number, proceed as with whole numbers, but place the decimal point in the quotient directly above the decimal point in the dividend.

EXAMPLE 2: Divide 20.46 by 66.

SOLUTION: Divide as with whole numbers, placing the decimal point in the quotient directly above the decimal point in the dividend. Check the answer by multiplying the quotient by the divisor.

$$
\begin{array}{r}
0.31 \\
66\overline{)20.46} \\
198 \\
\hline
66 \\
66 \\
\hline
0
\end{array}
$$

EXAMPLE 3: How many yards are in 165.6 inches?

SOLUTION: Since there are 36 inches in 1 yard, we divide the number of inches by 36, putting the decimal point in the quotient as in the previous example.

$$
\begin{array}{r}
4.6 \\
36\overline{)165.6} \\
144 \\
\hline
216 \\
216 \\
\hline
0
\end{array}
$$

To divide by a decimal, move the decimal point of the divisor to the right until it becomes a whole number. That is, multiply it by 10 or a multiple of 10. Next move the decimal point of the dividend the same number of places to the right, adding zeros if necessary. Multiplying the divisor and the dividend by the same number does not change the quotient.

EXAMPLE 4: Divide 131.88 by 4.2.

SOLUTION: Division of a decimal by a decimal is simplified if the divisor is made a whole number. In this case, the divisor 4.2 is made a whole number by moving the decimal point one place. Therefore, we move the decimal point one place in the dividend. Then we place the decimal point in the quotient directly above the decimal point in the dividend, and proceed as for division of whole numbers.

$$
\begin{array}{r}
31.4 \\
4{\underbrace{)2.}}\,{\overline{)131{\underbrace{8.8}}}} \\
126\phantom{.8} \\
58\phantom{.8} \\
42\phantom{.8} \\
168\phantom{} \\
168\phantom{} \\
0\phantom{}
\end{array}
$$

Check the answer by multiplying the quotient by the original divisor. The product must equal the original dividend. In this example,

$$31.4 \times 4.2 = 131.88.$$

To carry out a decimal quotient to a given number of places, add zeros to the right of the dividend until the dividend contains one more than the required number of places.

EXAMPLE 5: Find 0.3 ÷ 0.7 to the nearest thousandth.

SOLUTION: Often, division problems do not come out evenly. We then add zeros to the right of the dividend in order to carry out the division to the number of decimal places required.

$$
\begin{array}{r}
0.4285 \\
7{\overline{)3.0000}}
\end{array}
$$

## Estimation

As a general rule, carry out division to one more decimal place than is needed. If the last figure is 5 or more, drop it and add 1 to the figure in the preceding place. If the last figure is less than 5, just drop it completely.

## Averages

To find the **average** of several decimal quantities, divide their sum by the number of quantities.

EXAMPLE 6: Find the average of the following decimals: 1.734, 1.748, 1.64, and 1.802.

SOLUTION: Add the quantities and divide the sum, 6.924, by the number of quantities, or 4. Carry the answer to the thousandths place.

$$
\begin{array}{r}
1.734 \\
1.748 \\
1.640 \\
+\,1.802 \\
\hline
6.924
\end{array}
\qquad
\begin{array}{r}
1.731 \\
4{\overline{)6.924}}
\end{array}
$$

## Exercise Set 3.6

Divide, and find each answer to the nearest thousandth.

| | | | |
|---|---|---|---|
| **1.** | .34 ÷ 2 | **6.** | 1.11 ÷ .3 |
| **2.** | .35 ÷ 7 | **7.** | .987 ÷ 21 |
| **3.** | 5.4 ÷ 9 | **8.** | .2546 ÷ .38 |
| **4.** | 47.3 ÷ 10 | **9.** | 2.83 ÷ .007 |
| **5.** | 4.2 ÷ .01 | **10.** | .081 ÷ .0022 |

## Chapter 3 Glossary

**Average** The average of several quantities is the sum of the quantities divided by the number of quantities.

**Decimal Fraction** A fraction written as a multiple of 10, using a decimal point.

**Decimal Point** The period before the digits in a decimal.

## Chapter 3 Test

For each problem, five answers are given. Only one answer is correct. After you solve each problem, check the answer that agrees with your solution.

1. James has a metal frame with an 8.32-inch perimeter. If the sides are 0.04 inches thick, what is the perimeter of the inside of the frame?

   **A)** 8.16 inches    **D)** 8 inches
   **B)** 8.48 inches    **E)** 8.64 inches
   **C)** 8.24 inches

   *Note:* Perimeter is equal to the total distance around an object.

2. A safety pin is supposed to have a 0.675-inch diameter, but it was made 0.0007 inches too big. What is the diameter of the safety pin.

   **A)** 0.6743 inch    **D)** 0.4725 inch
   **B)** 0.6757 inch    **E)** 0.6675 inch
   **C)** 0.7675 inch

3. Tina sold pints of blueberries for $1.09 apiece at the Farmer's Market. If she earned $92.65, how many pints did she sell?

   **A)** 90 pints    **D)** 85 pints
   **B)** 100 pints    **E)** 75 pints
   **C)** 95 pints

4. A 16-story apartment building is 158.72 feet high. How high is the ceiling of the sixth floor from the ground?

   **A)** 9.92 feet    **D)** 59.52 feet
   **B)** 26.45 feet    **E)** 36.84 feet
   **C)** 66 feet

5. Rachel needs to cut 4 wire pieces that are each 6.42 feet long. How much wire does she need altogether?

   **A)** 32.75 feet    **D)** 33.45 feet
   **B)** 25.68 feet    **E)** 42.7 feet
   **C)** 24.08 feet

6. Write $^{17}/_{64}$ as a decimal.

   **A)** 0.2675    **D)** 0.275
   **B)** 0.276    **E)** 0.25564
   **C)** 0.265625

7. Write 0.375 as a fraction.

   **A)** $^{7}/_{16}$    **D)** $^{3}/_{8}$
   **B)** $^{5}/_{6}$    **E)** $^{2}/_{5}$
   **C)** $^{9}/_{14}$

8. Write 0.75 as a fraction.

   **A)** $^{3}/_{4}$    **D)** $^{7}/_{11}$
   **B)** $^{5}/_{14}$    **E)** $^{3}/_{8}$
   **C)** $^{6}/_{9}$

9. Write 0.1875 as a fraction.

   **A)** $^{4}/_{15}$    **D)** $^{14}/_{23}$
   **B)** $^{5}/_{25}$    **E)** $^{2}/_{7}$
   **C)** $^{3}/_{16}$

**10.** Write $^{79}/_{100}$ as a decimal.

    **A)** 0.398     **D)** 0.794
    **B)** 0.158     **E)** 0.368
    **C)** 0.79

**11.** Find the sum of 0.475, 0.279, and 1.456.

    **A)** 2.60     **D)** 2.75
    **B)** 3.546     **E)** 1.98
    **C)** 2.21

**12.** Find the difference between 3.79 and 2.81.

    **A)** 1.02     **D)** 0.92
    **B)** 0.78     **E)** 0.98
    **C)** 0.89

**13.** Find the sum of 2.79, 1.5789, 0.7847, and 0.274.

    **A)** 5.532     **D)** 5.379
    **B)** 5.4276     **E)** 5.6415
    **C)** 5.714

**14.** Find the difference between 0.9743 and 0.391.

    **A)** 0.5833     **D)** 0.624
    **B)** 0.6821     **E)** 0.59
    **C)** 0.5719

**15.** Multiply 0.043 by 1.79.

    **A)** 0.748     **D)** 0.0071
    **B)** 0.067     **E)** 0.08424
    **C)** 0.07697

**16.** Multiply 2.58 by 1.21.

    **A)** 3.1218     **D)** 3.762
    **B)** 3.044     **E)** 3.32
    **C)** 3.248

**17.** Divide 2.68 by 0.67.

    **A)** 4     **D)** 7
    **B)** 2     **E)** 5
    **C)** 11

**18.** Divide 3.75 by 0.25.

    **A)** 12     **D)** 21
    **B)** 9     **E)** 25
    **C)** 15

**19.** Divide 4.396 by 4.

    **A)** 1.97     **D)** 1.099
    **B)** 2.07     **E)** 1.054
    **C)** 1.22

**20.** Add 2.47 and 3.792, and then multiply the sum by 1.4.

    **A)** 8.7668     **D)** 8.9446
    **B)** 8.2277     **E)** 8.5449
    **C)** 8.679

**21.** Tim has 3 boards; one is 0.76 yards long, another is 0.648 yards long, and the third is 0.875 yards long. What is the average length of the 3 boards?

    **A)** 0.891     **D)** 0.761
    **B)** 0.788     **E)** 0.824
    **C)** 0.694

# Percents

## 4.1 Percents

**Percent,** or %, means a number of parts of one hundred (100). For example, 4% may be written as $\frac{4}{100}$ or 0.04. Notice that $\frac{4}{100}$ reduces to $\frac{1}{25}$. Percents may be added, subtracted, multiplied, or divided, just like other numbers.

EXAMPLE 1: Add 8% and 6%.

SOLUTION: $8\% + 6\% = 14\%$

EXAMPLE 2: Subtract 12% from 18%.

SOLUTION: $18\% - 12\% = 6\%$

EXAMPLE 3: Divide 18% by 9%.

SOLUTION: $18\% \div 9\% = 2$

EXAMPLE 4: Multiply 7% by 5.

SOLUTION: $7\% \times 5 = 35\%$

**Percentage** is a term used in arithmetic to denote that a whole quantity divided into a hundred equal parts is taken as the standard of measure.

The terms commonly used in percentage problems are rate ($R$), base ($B$), and percentage ($P$).

The **rate** ($R$) is the percent that is to be found. It is also called the **rate percent.** The **base** ($B$) is the whole quantity of which some percent is to be found. The percentage ($P$) is the result obtained by taking a given percent of the base.

EXAMPLE 5: What is 4% of 50?

SOLUTION: Here the rate is 4%, and the base is 50. Change 4% to a decimal fraction. Then multiply the decimal fraction, 0.04 by the base, 50.

$$0.04 \times 50 = 2$$

4% of 50 is 2.

## Exercise Set 4.1

1. Add 8% and 75%.
2. Subtract 12% from 56%.
3. Multiply 4 by 23%.
4. Divide 45% by 5.
5. Add 45% and 16%.
6. Add 14% and 26%.
7. Subtract 34% from 89%.
8. Subtract 11% from 49%.
9. Subtract 1% from 40%.
10. Multiply 21% by 12.
11. Multiply 6 by 66%.
12. Multiply 7% by 14.
13. Divide 49% by 7%.
14. Divide 49% by 7.
15. Divide 63% by 3.

# 4.2 Converting Percents to Decimals or Fractions

To change a percent to a decimal, remove the percent sign and move the decimal point two places to the left.

EXAMPLE 1: Change 25% to a decimal.

SOLUTION: Move the decimal point two places to the left.

$$25\% = 0.25$$

EXAMPLE 2: Change 1.5% to a decimal.

SOLUTION: To move the decimal point two places to the left, one zero needs to be put in front of 15.

$$1.5\% = 0.015$$

To change a percent to a fraction, divide the percent quantity by 100, and reduce to lowest terms.

EXAMPLE 3: Convert 8% to a fraction in lowest terms.

SOLUTION: $8\% = {}^8/_{100} = {}^2/_{25}$    $= 0.08$

EXAMPLE 4: Convert 75% to a fraction in lowest terms.

SOLUTION: $75\% = {}^{75}/_{100} = {}^3/_4$    $= 0.75$

EXAMPLE 5: Convert 80% to a fraction in lowest terms.

SOLUTION: $80\% = {}^{80}/_{100} = {}^4/_5$    $0.8$

### FRACTIONAL EQUIVALENTS OF PERCENTS

| | | |
|---|---|---|
| $10\% = {}^1/_{10}$ | $12\tfrac{1}{2}\% = {}^1/_8$ | $8\tfrac{1}{3}\% = {}^1/_{12}$ |
| $20\% = {}^1/_5$ | $25\ \% = {}^1/_4$ | $16\tfrac{2}{3}\% = {}^1/_6$ |
| $40\% = {}^2/_5$ | $37\tfrac{1}{2}\% = {}^3/_8$ | $33\tfrac{1}{3}\% = {}^1/_3$ |
| $50\% = {}^1/_2$ | $62\tfrac{1}{2}\% = {}^5/_8$ | $66\tfrac{2}{3}\% = {}^2/_3$ |
| $60\% = {}^3/_5$ | $87\tfrac{1}{2}\% = {}^7/_8$ | $83\tfrac{1}{3}\% = {}^5/_6$ |

## Exercise Set 4.2

Convert the percentages to fractions in lowest terms.

1. 1% =
2. 2% =
3. 4% =
4. 7% =
5. ½% =
6. 6¼% =
7. 6⅔% =
8. 7½% =
9. ⅓% =
10. ¾% =
11. 1½% =
12. 3½% =

Convert the percentages to decimals.

**13.** 42% =
**14.** 16½% =
**15.** 231% =
**16.** 1% =
**17.** ¼% =
**18.** 11% =

**19.** 38% =
**20.** 67% =
**21.** 4% =
**22.** 44% =
**23.** 21²⁄₄% =
**24.** 50³⁄₈% =

## 4.3 Converting Decimals to Percents

To change a decimal to a percent, move the decimal point two places to the right and add a percent sign.

EXAMPLE 1: Change 0.24 to a percent.

SOLUTION: Move the decimal point two places to the right and add the % sign.

$$0.24 = 24\%$$

EXAMPLE 2: Change 0.0043 to a percent.

SOLUTION: Note that this is less than 1%.

$$0.0043 = 0.43\%$$

EXAMPLE 3: Change 2.45 to a percent.

SOLUTION: Note that any whole number greater than 1 which designates a percent is more than 100%.

$$2.45 = 245\%$$

### Exercise Set 4.3

Change each decimal to a percent.

**1.** 0.436 =
**2.** 0.21 =
**3.** 4.32 =
**4.** 0.99 =

**5.** 0.740 =
**6.** 2.15 =
**7.** 0.11 =
**8.** 0.68 =
**9.** 0.875 =
**10.** 0.75 =

**11.** 0.325 =
**12.** 5.675 =
**13.** 19.895 =
**14.** 2.531 =
**15.** 50.249 =

## 4.4 Percentage Problems

The first type of percentage problem we will discuss involves finding the percent of a number, given the base and the rate. The formula we will use is:

Percentage = Base × Rate; $P = B \times R$

EXAMPLE 1: Find 14% of $300.

SOLUTION: The rate is 14%. The base is $300. To find the percentage, multiply $300 by 0.14.

$$14\% = 0.14$$
$$\$300 \times 0.14 = \$42$$

14% of $300 is $42

The second type of percentage problem involves finding what percent one number is of another, given the base and the percentage. The formula is:

Rate = Percentage ÷ Base; $R = P \div B$

EXAMPLE 2: What percent of 240 is 120?

SOLUTION: The base is 240. The percentage is 120. Divide the percentage by the base to get the rate.

$$^{120}/_{240} = ½$$

Rewrite the answer, $\frac{1}{2}$, as the percent, 50%. This is another way of saying what fractional part of 240 is 120. So, 120 is 50% of 240.

The third type of percentage problem involves finding the base when the rate and the percentage of the base are known. The formula is:

Base = Percentage ÷ Rate; $B = P \div R$

EXAMPLE 3: 225 is 25% of what amount?

SOLUTION: This question is another way of saying that $\frac{1}{4}$ of some number equals 225, and what is that whole number?

The percentage is 225. The rate is 25%. Divide the percentage by the rate to find the base.

$$25\% = \frac{25}{100} = \frac{1}{4}$$

$$225 \div \frac{1}{4} = 225 \times 4 = 900$$

The base is 900.

These formulas make the solution of percentage problems simpler, if we learn to identify the base, the rate, and the percentage.

## Exercise Set 4.4

1. Find 30% of 620 gallons.
2. Find $12\frac{1}{2}$% of 96 men.
3. What is 4% of 250 pounds?
4. How much is $62\frac{1}{2}$% of $80?
5. Find $\frac{1}{2}$ of 1% of 190 tons.
6. What percent of 32¢ is 8¢?
7. What percent of 15 inches is $7\frac{1}{2}$ inches?
8. What percent of $200 is $14?
9. What percent of $\frac{2}{3}$ is $\frac{1}{3}$?

10. What percent of $\frac{2}{3}$ is $\frac{1}{2}$?
11. Twelve is 25% of what number?
12. Ten is 20% of what number?
13. Eight is $2\frac{1}{2}$% of what number?
14. Sixteen percent of a sum is 128. What is the sum?
15. What number increased by 25% of itself equals 120?

## 4.5 Ratios and Proportions

A **ratio** is the relation between two like numbers, or two like values. The ratio may be written as a fraction, $\frac{3}{4}$, as a division, $3 \div 4$, or with the colon or ratio sign (:), $3:4$. When the last of these forms is used, it is read, "3 to 4," or "3 is to 4." Ratios may be expressed by the word per as in miles per hour. This ratio may be written miles/hour. However, we write the ratio, its value in arithmetical computations always remains the same.

Since a ratio may be regarded as a fraction, multiplying or dividing both terms of a ratio by the same number does not change the value of the ratio.

Thus, $\frac{2}{4} = \frac{4}{8}$, as $\frac{4}{8}$ is obtained by multiplying both terms in $\frac{2}{4}$ by 2.

To reduce a ratio to its lowest terms, treat the ratio as a fraction and reduce the fraction to its lowest terms.

EXAMPLE 1: Express the ratio $\frac{2}{3}$ to $\frac{4}{9}$ in lowest terms.

SOLUTION:
$\frac{2}{3}$ to $\frac{4}{9} = \frac{2}{3} \div \frac{4}{9} = \frac{2}{3} \times \frac{9}{4} = \frac{3}{2}$.

Thus $\frac{2}{3}$ to $\frac{4}{9}$ can be rewritten as $\frac{3}{2}$, or as 3 to 2.

To separate a quantity according to a given ratio, add the terms of the ratio to find the total number of parts. Find what fractional part each term is of the whole. Divide the total quantity into parts corresponding to the fractional parts.

EXAMPLE 2: Three hundred tents have to be divided between 2 Scout troops in the ratio of 1 to 2. How many tents does each troop get?

SOLUTION: First, add the terms of the ratio.

$$1 + 2 = 3$$

Now take corresponding fractional parts of the total quantity to be divided.

$$\frac{1}{3} \times 300 = 100$$

$$\frac{2}{3} \times 300 = 200$$

*Check:* $100:200 = {}^{100}\!/_{200} = \frac{1}{2}$, or $1:2$.

One Scout troop gets 100 tents, and the second Scout troop gets 200 tents.

EXAMPLE 3: Three dealers receive 1,600 pounds of coffee in the ratio of 8:11:13. How many pounds should each dealer get?

SOLUTION: First add the terms of the ratio.

$$8 + 11 + 13 = 32$$

$$\frac{8}{32}, \frac{11}{32}, \frac{13}{32} \text{ are the fractional parts}$$

$$\frac{8}{32} \times 1,600 = 400$$

$$\frac{11}{32} \times 1,600 = 550$$

$$\frac{13}{32} \times 1,600 = 650$$

The first dealer gets 400 pounds of coffee, the second dealer gets 550 pounds of coffee, and the third dealer gets 650 pounds of coffee.

A **proportion** is a statement of equality between two ratios. It may be written with the double colon or **proportion sign** (::), or with the sign of equality (=).

Thus, 2:6 :: 3:9 is a proportion that is read, 2 is to 6 as 3 is to 9; or $\frac{2}{6}$ equal $\frac{3}{9}$.

In any proportion the first and last terms are called the **extremes** and the second and third terms are called the **means.** In 2:6 :: 3:9 the extremes are 2 to 9; and the means are 6 and 3.

Multiply the two extremes and the two means of the proportion 2:6 :: 3:9 and compare the products.

Extremes: $2 \times 9 = 18$

Means:   $6 \times 3 = 18$

| The product of the means is equal to the product of the extremes. |
| --- |

If we write the proportion in the form $\frac{2}{6} = \frac{3}{9}$, note that the means and extremes are diagonally opposite one another.

No proportion is a true proportion unless the two ratios are equal. We can find the missing term of any proportion, given three of the terms, by using the rule stating that the product of the means equals the product of the extremes.

EXAMPLE 4: Find the value of the missing term in the proportion 2:6 = 8:?

SOLUTION: The letter $x$ is traditionally used to denote a missing term or an unknown quantity. Rewrite the proportion.

2:6 :: 8:$x$.

2 times $x$ = 6 times 8    Product of the extremes equals product of the means.

$$2x = 48$$

$$\frac{2x}{2} = \frac{48}{2}$$    Both sides of any equation may be divided by the same number without changing the equation.

$$x = 24$$

This process is the equation method of solving problems containing an unknown. We will learn more about this process in later chapters.

If we wish to use a strict arithmetic method of finding the missing term in a proportion, we can try either of two methods.

---

**Method 1**

The product of the means divided by either extreme gives the other extreme as the quotient.

---

EXAMPLE 5: Find the value of the missing extreme in the proportion ?:6 = 8:24.

SOLUTION: First, multiply the two means.

$$6 \times 8 = 48$$

Divide this product by the known extreme.

$$48 \div 24 = 2$$

The quotient 2 is the unknown extreme.

---

**Method 2**

The product of the extremes divided by either mean gives the other mean as a quotient.

---

EXAMPLE 6: Find the value of the missing mean in the proportion

3:? = 8:24.

SOLUTION: Multiply the two extremes.

$$3 \times 24 = 72$$

Now divide this product by the given mean.

$$72 \div 8 = 9$$

The quotient 9 is the missing mean.

When solving problems by the ratio and proportion method it is first necessary to recognize whether a proportion exists, and if so what kind it is.

A **direct proportion** is indicated when two quantities are related so that an increase in one causes a corresponding increase in the other or when a decrease in one causes a corresponding decrease in the other.

The list following shows typical quantitative expressions in which two variables are directly related.

a. The faster the speed, the greater the distance covered.
b. The more men working, the greater the amount of work done.
c. The faster the speed, the greater the number of revolutions.
d. The higher the temperature of gas, the greater the volume.
e. The taller the object, the longer the shadow.
f. The larger the quantity, the greater the cost.
g. The smaller the quantity, the lower the cost.
h. The greater the length, the greater the area.
i. The greater the base, the larger the discount, commission, interest, and profit.

EXAMPLE 7: If 20 men assemble 8 cars in a day, how many men are needed to assemble 12 cars in a day?

SOLUTION: We make a proportion, since the quantities are directly related. The more cars assembled, the more men needed. Solve for $x$.

$$8 \text{ cars need } 20 \text{ men}$$

$$12 \text{ cars need } ? \text{ men}$$

$$8 : 12 :: 20 : x$$

$$8x = 240$$

$$x = 30$$

30 men are needed to assemble twelve cars in one day

EXAMPLE 8: If 12 pads of paper cost $8.00, how much will 9 pads of paper cost?

SOLUTION: The values are in direct proportion, as the fewer the number of pads, the lower the cost. Solve for $x$.

$$12 \text{ pads cost } \$8.00$$

$$9 \text{ pads cost } ?$$

$$12 : 9 :: 8 : x$$

$$12x = 72$$

$$x = \$6.00$$

9 pads of paper cost $6.00
In every proportion both ratios must be written in the same order of value, for instance in EXAMPLE 2:

$$\frac{\text{smaller no. of cars}}{\text{larger no. of cars}} = \frac{\text{smaller no. of men}}{\text{larger no. of men}}$$

In EXAMPLE 8:

$$\frac{\text{larger no. of pads}}{\text{smaller no. of pads}} = \frac{\text{larger cost}}{\text{smaller cost}}$$

An **inverse proportion** is indicated when two quantities are related so that an increase in one causes a corresponding decrease in the other, or vice versa.
The following list shows quantitative expressions in which two variables are inversely related.

a. The greater the speed, the less the time.
b. The slower the speed, the longer the time.
c. The greater the volume, the less the density.
d. The more men working, the shorter the time.

PRACTICALLY SPEAKING 4.5

Martha is flying to England for a visit. She intends to exchange her U.S. dollars for English pounds. The exchange rate allows 60 pounds for every 100 U.S. dollars.

1. How many pounds will she get if she exchanges $450 U.S. dollars for pounds?

See Appendix F for the answer.

---

e. The fewer men working, the longer the time.

EXAMPLE 9: When 2 pulleys are belted together, the revolutions per minute, or rpms, vary inversely with the size of the pulleys. A 20-inch pulley running at 180 rpm drives an 8-inch pulley. Find the revolutions per minute of the 8-inch pulley.

SOLUTION: First, make a table of corresponding values. Put like quantities together. The quantities are in inverse proportion, so the smaller the pulley, the greater the number of revolutions. Invert the first ratio and write the proportion. Solve for $x$.

20-inch pulley makes 180 rpm

8-inch pulley makes ? rpm

$$\frac{8}{20} = \frac{180}{x}$$

$$8x = 3,600$$

$$x = 450 \text{ rpm}$$

Note that when we write the inverse proportion from EXAMPLE 9 in the form $\frac{8}{20} = \frac{180}{x}$ the corresponding numbers are arranged diagonally. That is, 20 inches is diagonally opposite 180 rpm, and 8 inches is diagonally opposite $x$ rpm.

In the direct proportion from EXAMPLE 8, $\frac{12}{9} = \frac{8}{x}$, the corresponding numbers are arranged directly in line with one another. That is 12 pads is across from $8.00, and 9 pads is across from $6.00.

## Exercise Set 4.5

Reduce each ratio to lowest terms.

1. $25:30$
2. $\frac{6}{27}$
3. $14:84$
4. $\frac{3}{53}$
5. $11:121$
6. $\frac{3}{9}:\frac{4}{7}$
7. $\frac{16}{4}$ to $\frac{11}{3}$
8. $\frac{1}{4}:\frac{6}{10}$
9. $\frac{33}{600}$
10. $12:14$
11. $78:4$

Solve each problem.

12. Ted and Tasha held a tag sale to get rid of some of their unused belongings. The profits totaled $325, and they split the profits using a 40:60 ratio. How much money did each get?

13. A casserole recipe requires 2 eggs for every 3 people served. If Jeff makes the recipe for 9 people, how many eggs does he need?

14. Jenny's lamp is made of a metal that contains 8% tin. If the lamp weighs 72 ounces, how many ounces of tin are contained in the lamp?

15. Alan has 36 apples that he must arrange in 2 baskets. If the first basket will hold twice as much as the second basket, how many apples will go in each basket?

Find the missing term in each proportion.

| | |
|---|---|
| **16.** 2:3 :: 4:? | **21.** 5:? :: 25:20 |
| **17.** 20:10 = ?:6 | **22.** ?:5 :: 12:20 |
| **18.** 2:? = 8:24 | **23.** ?:25 = 10:2 |
| **19.** 18:? = 36:4 | **24.** 9:? :: 24:8 |
| **20.** 12:4 = ?:7 | **25.** 24:4 = ?:3 |

Solve each problem using proportions.

26. A backpacking organization has enough food packed for 240 people for 28 days. However, only 112 people go on this backpacking trip. How long will the food last?

27. A train traveling at 35 mph takes 26 hours to travel from Chicago to New York. How fast must the train travel to make the trip in 20 hours?

28. The flywheel on an engine makes 220 revolutions in 2 seconds. How many revolutions does the flywheel make in 8 seconds?

# Chapter 4 Glossary

**Base**   The whole quantity of which some percent is to be found.

**Direct Proportion**   Two quantities related so that an increase in one causes an increase in the other or when a decrease in one causes a decrease in the other.

**Extremes**   The first and last terms in any proportion.

**Inverse Proportion**   Two quantities related so that an increase in one causes a decrease in the other or when a decrease in one causes an increase in the other.

**Means**   The second and third terms in any proportion.

**Percent**   The number of parts per hundred.

**Percentage**   The result obtained by taking a given percent of the base.

**Proportion**   The statement of equality between two ratios.

**Rate**   The fractional percent in hundredths that is to be found.

**Ratio**   The relation between two like numbers, or values.

# Chapter 4 Test

For each problem, five answers are given. Only one answer is correct. After you solve each problem, check the answer that agrees with your solution.

1. Write $\frac{6}{100}$ as a percent.

| | | | |
|---|---|---|---|
| **A)** 60% | | **D)** 30% |
| **B)** 3% | | **E)** 12% |
| **C)** 6% | | |

**2.** What is 30% of 620 gallons?

    **A)** 186 gallons    **D)** 199 gallons
    **B)** 215 gallons    **E)** 178 gallons
    **C)** 220 gallons

**3.** Find $12\frac{1}{2}\%$ of 96.

    **A)** 16    **D)** 12
    **B)** 8    **E)** 25
    **C)** 50

**4.** What is 4% of 250 pounds?

    **A)** 16 pounds    **D)** 45 pounds
    **B)** 25 pounds    **E)** 10 pounds
    **C)** 12 pounds

**5.** How much is $62\frac{1}{2}\%$ of $80?

    **A)** $62    **D)** $48
    **B)** $34    **E)** $25
    **C)** $50

**6.** What percent of $32 is $8.00?

    **A)** $12\frac{1}{2}\%$    **D)** 8%
    **B)** 50%    **E)** 16%
    **C)** 25%

**7.** What percent of 15 inches is $7\frac{1}{2}$ inches?

    **A)** 12%    **D)** 8%
    **B)** 50%    **E)** 75%
    **C)** 19%

**8.** What percent of $200 is $14?

    **A)** 30%    **D)** 20%
    **B)** 17%    **E)** 7%
    **C)** 15%

**9.** What percent of $\frac{2}{3}$ is $\frac{1}{3}$?

    **A)** 25%    **D)** 15%
    **B)** 35%    **E)** 10%
    **C)** 50%

**10.** What percent of $\frac{2}{3}$ is $\frac{1}{2}$?

    **A)** 45%    **D)** 25%
    **B)** 90%    **E)** 75%
    **C)** 50%

**11.** Twelve is 25% of what number?

    **A)** 48    **D)** 36
    **B)** 24    **E)** 56
    **C)** 60

**12.** Ten is 20% of what number?

    **A)** 20    **D)** 50
    **B)** 100    **E)** 120
    **C)** 45

**13.** Eight is $2\frac{1}{2}\%$ of what number?

    **A)** 450    **D)** 320
    **B)** 375    **E)** 250
    **C)** 355

**14.** What number is 16% of 128?

    **A)** 25    **D)** 29.276
    **B)** 28.4    **E)** 48.26
    **C)** 20.48

**15.** What number increased by 25% of itself equals 120?

    **A)** 70    **D)** 90
    **B)** 96    **E)** 85
    **C)** 102

**16.** A loaded 18-wheeler weighs 20,000 pounds. If 80% of this represents the load, how much does just the truck weigh?

    **A)** 2,000 pounds
    **B)** 8,000 pounds
    **C)** 4,000 pounds
    **D)** 16,000 pounds
    **E)** 12,000 pounds

**17.** A garbage dumpster weighs 8% as much as its contents. If the garbage weighs 275 pounds, what is the weight of the garbage dumpster?

    **A)** 27 pounds     **D)** 22 pounds
    **B)** 65 pounds     **E)** 35 pounds
    **C)** 14 pounds

**18.** A brass lamp weighing 75 pounds is constructed of 45% zinc, with the balance of the metal being copper. How many pounds of copper are contained in the lamp?

    **A)** 55 pounds     **D)** $12\frac{3}{5}$ pounds
    **B)** $41\frac{1}{4}$ pounds     **E)** 25 pounds
    **C)** $33\frac{3}{4}$ pounds

**19.** What number increased by 75% of itself equals 140?

    **A)** 80     **D)** 75
    **B)** 96     **E)** 105
    **C)** 90

**20.** A truck carrying 6,750 pounds of coal weighed 9,000 pounds. What percent of the total weight was due to the weight of the truck?

    **A)** 15%     **D)** 30%
    **B)** 20%     **E)** 35%
    **C)** 25%

**21.** A bronze statue with a tin base weighed 28 pounds. If the base weighs $3\frac{1}{2}$ pounds, what percent of the total weight is bronze?

    **A)** $87\frac{1}{2}$%     **D)** $83\frac{1}{3}$%
    **B)** $12\frac{1}{2}$%     **E)** 95%
    **C)** 9.8%

**22.** In Gordon's Health Club, 8% of the exercise machines broke down. How many exercise machines are there altogether if 32 machines have to be repaired?

    **A)** 250 exercise machines
    **B)** 620 exercise machines
    **C)** 560 exercise machines
    **D)** 400 exercise machines
    **E)** 380 exercise machines

**23.** If a pole 18 feet high casts a shadow 20 feet long, how long a shadow will a pole 27 feet high cast?

    **A)** 10 feet     **D)** 36 feet
    **B)** 25 feet     **E)** 21 feet
    **C)** 30 feet

**24.** If Jane walks 9 miles in 2 hours, how long will it take her to walk 30 miles?

    **A)** 6 hours     **D)** 9 hours
    **B)** $6\frac{2}{3}$ hours     **E)** 4 hours
    **C)** $8\frac{1}{2}$ hours

**25.** If a jeep runs 90 miles on 5 gallons of gas, how far will it run on a full 20-gallon tank?

    **A)** 300 miles     **D)** 280 miles
    **B)** 360 miles     **E)** 420 miles
    **C)** 450 miles

# Signed Numbers

## 5.1 Signed Numbers

Up until now, all the numbers used have been positive numbers. That is, none was less than zero (0). In solving some problems in arithmetic it is necessary to assign a negative value to some numbers. This is used principally for numbers with which we wish to represent opposite quantities that can best be illustrated by use of a diagram. For example consider a thermometer, as in Figure 1.

**Figure 1.**

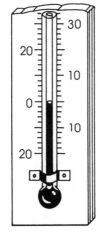

If temperatures above zero are taken as positive, then temperatures below zero are considered negative.

In measuring distances east and west, as in Figure 2, if distance east of a certain point is taken as positive, then distance west of that point is considered negative.

**Figure 2.**

Another good example may be taken from bookkeeping, where money in the bank may be considered a positive amount, while money owed represents a negative amount.

Thus, positive and negative numbers are used to distinguish between opposite qualities. Values above zero are considered positive and take the + sign, while values below zero are considered negative and are written with the − sign. These are called signed numbers.

The + and − also continue to be used as signs of addition and subtraction. When no sign is indicated the + sign is understood to be used.

Learning to use signed numbers is an introduction to some of the special rules for algebraic operations.

The set of whole numbers along with their opposites is called the set of **integers,** or $\{. . . , -3, -2, -1, 0, 1, 2, 3, . . .\}$.

## 5.2 Adding and Subtracting Signed Numbers

To add numbers with like signs, add the numbers as in arithmetic, and give the result the common sign.

EXAMPLE 1: Add $-14$ and $-8$.

SOLUTION: If we add positive numbers, then $14 + 8 = 22$. The common sign is negative, so

$$-14 + (-8) = -22.$$

EXAMPLE 2: Add $+4$, $+12$, and $+16$.

SOLUTION: If we add positive numbers, then $4 + 12 + 16 = 32$. The common sign is positive, so

$$+4 + (+12) + (+16) = +32$$

To add numbers with unlike signs, combine all positive and negative quantities, subtract the smaller from the larger, and give the result the sign of the larger combination.

EXAMPLE 3: Add $-4 + (-8) + 2 + 6 + 10$.

SOLUTION: First, add together numbers with the same sign.

$$-4 + (-8) = -12$$
$$2 + 6 + 10 = 18$$

Now subtract the absolute value of the smaller number from the absolute value of the larger number.

$$18 - 12 = 6$$

The answer takes the sign of the larger number, so the answer is $+6$.

EXAMPLE 4: Add $3 + 19 + 4 + (-45)$.

SOLUTION: First add together numbers with the same signs.

$$3 + 19 + 4 = \quad 26$$
$$-45 \qquad = -45$$

Now subtract the larger number from the smaller number.

$$45 - 26 = 19$$

Since the sign of the larger number is negative, the answer is $-19$.

Subtraction means finding the difference between two numbers. If we were asked what the difference was between $-4°$ centigrade and $+5°$, the answer would be $9°$. We can do this mentally. Now how do we arrive at this answer? First, we count from $-4°$ to zero, and then add 5 to that.

Therefore, to subtract signed numbers, we change the sign of the subtrahend, and apply the rules for addition.

---

**PRACTICALLY SPEAKING 5.2**

Renée researches stories for the Weather Channel. She is gathering information about extreme temperature changes in Nevada. Renée sees a change one day from $-17°$F. at midnight to $105°$F. at noon.

1. What is the total temperature change in that 12-hour period?

See Appendix F for the answer.

---

EXAMPLE 5: Subtract $+20$ from $+32$.

SOLUTION: Since $+20$ is the subtrahend, or the number to be subtracted, we change the sign and add.

$$32 + (-20) = 12$$

EXAMPLE 6: Subtract $-12$ from $-18$.

SOLUTION: Since $-12$ is the subtrahend, we change the sign of $-12$, and add.

$$-18 + (+12) = -6$$

## Exercise Set 5.2

Add or subtract as indicated.

1. $5 + 18 =$
2. $-5 + (-17) + (-14) =$
3. $+7 + (-12) - 6 + 4 =$
4. $22 - 14 - 17 - 12 + 18 =$
5. $47 - 19$
6. $-26 - (-17) =$
7. $-42 + 18 =$
8. $54 - (-12) =$
9. $80 - (-50) =$
10. $-5 - (-8) =$
11. $-7 - (-4) =$
12. $-9 - 16 =$
13. $233 - (-47) =$
14. $-72 + (-69) =$
15. $-45 + (-115) =$

## 5.3 Multiplication and Division of Signed Numbers

When multiplying signed numbers, the product of any two numbers that have like signs is positive $(+)$, and the product of any two numbers that have unlike signs is negative $(-)$.

EXAMPLE 1: Multiply $-8$ by $-6$.

SOLUTION: Since the signs are the same, the product is positive.

$$-8 \times (-6) = +48$$

EXAMPLE 2: Multiply $+3$ by $-4$.

SOLUTION: Since the signs are unlike, the product is negative.

$$+3 \times (-4) = -12$$

EXAMPLE 3: Multiply $-2$ by 5 by $-3$ by 4.

SOLUTION:

$$
\begin{aligned}
-2 \times 5 \times (-3) \times 4 &= (-2 \times 5) \times -3 \times 4 \\
&= -10 \times -3 \times 4 \\
&= (-10 \times -3) \times 4 \\
&= 30 \times 4 \\
&= 120
\end{aligned}
$$

Division of signed numbers is carried out by the same process as division in arithmetic. However, the sign of the quotient is positive if the divisor and the dividend have the same sign, and negative if the divisor and dividend have opposite signs.

EXAMPLE 4: Divide $-16$ by $-2$.

SOLUTION: Since the signs of both the dividend and the divisor are the same, the quotient is positive.

$$-16 \div -2 = +8$$

EXAMPLE 5: Divide $-35$ by $+5$.

SOLUTION: Since the signs of the divisor and the dividend are different, the quotient is negative.

$$-35 \div 5 = -7$$

## Exercise Set 5.3

Simplify the following expressions.

1. $2 \times (-16) =$
2. $(-18) \times (-12) =$
3. $(-4) \times (-6) \times 3 =$
4. $4 \times 3 \times (-2) \times 6 =$
5. $72 \div (-24) =$
6. $(-68) \div (-17) =$
7. $(-14) \div (-5) =$
8. $-24 \times 4 \div 8 =$
9. $12 \div 3 \times (-16) =$
10. $8 \times (-3) \div (-6) =$
11. $16 \times (-4) \div 3 =$
12. $112 \times 2 \div (-42) =$
13. $(-64) \div (-8) \times 15 =$

14. $(-2) \times 45 \div 12 =$
15. $3 \times 14 \div 2 =$
16. $21 \div (-7) \times 14 =$
17. $222 \div 2 \times 7 =$
18. $52 \div -13 \times 15 =$
19. $3 \times -6 \times 12 \div 36 =$
20. $77 \div 11 \times 13 =$

## 5.4 Order of Operations

When we simplify expressions with more than one operation to perform, there is an order of priorities to follow.

---

**Order of Operations**

1) Perform operations inside of parentheses first. If there is more than one set of parentheses, start with the innermost set.
2) Evaluate powers and roots in order from left to right.
3) Perform all multiplications or divisions in order from left to right.
4) Perform all additions or subtractions in order from left to right.

---

**Parentheses** () or braces {} mean that the quantities inside are to be grouped together, and that the quantities enclosed are to be considered as one quantity. The line of a fraction has the same significance in this respect as a pair of parentheses.

Thus, $18 + (9 - 6)$ is read 18 plus the quantity $9 - 6$.

To solve problems containing parentheses, do the work inside the parentheses

BODMAS
(Bracket) = Div Mul Add Sub
BEDMAS
or
$2^2$

---

**PRACTICALLY SPEAKING 5.4**

Jim has $15 and he needs to buy food for dinner for himself and his wife. He stops at Food Mart to buy:

2 steaks @ $3.50 each

1 can of mushrooms @ $1.15

2 onions @ $0.39 each

1 loaf of bread @ $2.49

1 package of chocolate fudge @ $4.59

1. Jim needs to know if he has enough money to buy all of these items. He estimates the cost by rounding the price of each of the items to the nearest 0.50. Using this rounded figure, will Jim have enough money to buy everything he wants? What is the estimated price?

2. If Jim estimates the cost by rounding the price of each of the items to the nearest $0.10, will he believe he has enough money? What is the estimated price?

3. What items can Jim buy with the money he has with him? What factors would influence his decision?

See Appendix F for the answers.

---

first. Then remove the parentheses, and proceed with the other operations.

Note that two numbers separated only by parentheses like 2(4) are multiplied together. Therefore, 2(4) = 8.

It is extremely important to observe this method of procedure, since it is otherwise impossible to solve algebraic problems correctly.

EXAMPLE 1: Simplify 94 − (12 + 18 + 20).

SOLUTION:          94 − 50 = 44

EXAMPLE 2: Simplify 12(3 + 2).

SOLUTION:          12 × 5 = 60

EXAMPLE 3: Simplify $\dfrac{18}{2(4 - 1)}$.

SOLUTION:     $\dfrac{18}{2 \times 3} = \dfrac{18}{6} = 3$

EXAMPLE 4: Simplify 3 × 6 − 4.

SOLUTION:          18 − 4 = 14

Keep in mind that multiplications and divisions have equal priority. That is, when evaluating an expression, either a multiplication or a division may be performed first.

EXAMPLE 5: Evaluate $28 \div 14 \times 34$.

SOLUTION: Since there are no parentheses, powers, or roots to simplify, we perform all multiplications and divisions in order from left to right.

$$28 \div 14 \times 34 = 2 \times 34$$
$$= 68$$

Additions and subtractions also have equal priority, so either one may be performed first, from left to right.

EXAMPLE 6: Evaluate $34 - 21 + 45 + 11$.

SOLUTION: Since there are no parentheses, powers, roots, multiplications, and divisions, we perform all additions and subtractions in order from left to right.

$$34 - 21 + 45 + 11 = 13 + 45 + 11$$
$$= 58 + 11$$
$$= 69$$

EXAMPLE 7: Evaluate $(23 \times 4 + 17) + (25 - 16)$.

SOLUTION: Begin by simplifying the expressions within parentheses.

$$(23 \times 4 + 17) + (25 - 16)$$
$$= (92 + 17) + (9)$$
$$= 109 + 9$$
$$= 118$$

## Exercise Set 5.4

Evaluate each expression.

1. $(23 - 5) \div (3 + 12)$
2. $2 \times 3 - 14$
3. $16 \div 4 \times 27$
4. $121 \div 11 \times 49$
5. $35 \times 6 \div 25$
6. $99 \div 3 \times 16 \div 4$
7. $51 \div 3 \times 55 \times 2$
8. $42 \times 7 \div 3$
9. $45 \div 9 \times 33$
10. $83 \times 17 \div 5$

Clear parentheses or braces and solve.

11. $18 + (19 - 14) =$
12. $22(3 + 2) =$
13. $42 - 9 - (18 + 2) =$
14. $(6 - 4)(8 + 2) =$
15. $(18 \div 3)(9 - 7) =$
16. $(7 \times 8) - (6 \times 4) + (18 - 6) =$
17. $(6 \times 8) \div (8 \times 2) =$
18. $19 + (18 - 14 + 32) =$
19. $(7 \times 6)(6 \times 5) =$
20. $69 \div \{35 - (15 - 3)\} =$

## 5.5 Absolute Value

Now that we have defined the set of integers, we can discuss **absolute value.** The absolute value of any number is the distance of the number from zero. Absolute value is symbolized by putting the number or expression inside these two upright bars, $||$. The absolute value of a number is always positive, as absolute value measures the distance a number lies from zero.

$$|x| = x \text{ if } x \text{ is a positive number, or zero.}$$
$$|x| = -x \quad \text{if } x \text{ is negative}$$

EXAMPLE 1: Evaluate $|-15|$.

SOLUTION: $|-15| = 15$

Number lines are frequently used to illustrate the idea of absolute value, and to help show that distance is positive. Number lines list numbers from left to right, smaller to larger. This number line shows that the distance between $-5$ and zero is five units.

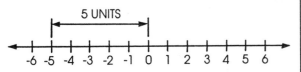

EXAMPLE 2: Use a number line to evaluate $|+7|$.

SOLUTION: Seven is seven units from zero.

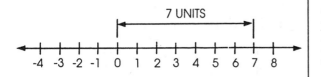

We can find the absolute value of an expression, as well as the absolute value of a number.

EXAMPLE 3: Evaluate $|25 - 69|$.

SOLUTION: First, find the result of the subtraction of the expression inside absolute value bars.

$$25 - 69 = -44$$

The expression simplifies to $|-44|$. Since we want the absolute value of the expression, we find the answer is 44. The distance between 25 and 69 is 44 units.

## Exercise Set 5.5

Evaluate each expression to find the absolute value.

1. $|-57|$
2. $|275|$
3. $|76 - 12|$
4. $|65 - 89|$
5. $|26 - 11|$
6. $|405 + 17 - 231|$
7. $|37 - 45 \times 11|$
8. $|55 \div 11 + 25|$
9. $|16 + 34 \times 3|$
10. $|25 \times 3 - 19|$

## Chapter 5 Glossary

**Absolute Value** The distance of any number from zero.

**Integers** The set of whole numbers along with their opposites.

**Negative Numbers** Numbers less than zero.

**Order of Operations** 1. Perform operations inside parentheses or brackets first. 2. Evaluate powers or roots from left to right. 3. Perform all multiplications or divisions from left to right. 4. Perform all additions or subtractions from left to right.

**Parentheses** Enclosures for quantities to be grouped together.

**Positive Numbers** Numbers greater than zero.

## Chapter 5 Test

For each problem, five answers are given. Only one answer is correct. After you solve each problem, check the answer that agrees with your solution.

1. Find the sum of $-14, 16, 21,$ and $-35$.

   **A)** 12      **D)** 16
   **B)** 75      **E)** $-32$
   **C)** $-12$

2. Find the difference between $-68$ and 44.

   **A)** 14      **D)** $-112$
   **B)** 112     **E)** $-24$
   **C)** 42

3. Find the sum of $72, -23,$ and $-84$.

   **A)** $-110$    **D)** 140
   **B)** $-45$     **E)** 45
   **C)** $-35$

4. Find the difference between $-144$ and 210.

   **A)** 354      **D)** 154
   **B)** $-75$    **E)** 825
   **C)** $-265$

5. Find the difference between 268 and $-142$.

   **A)** 720      **D)** 116
   **B)** 410      **E)** 390
   **C)** 240

6. Multiply 25 by $-31$.

   **A)** 850      **D)** $-605$
   **B)** $-775$   **E)** 775
   **C)** 605

7. Find the product of $-13$ and $-16$.

   **A)** 208      **D)** 178
   **B)** 240      **E)** $-195$
   **C)** $-208$

8. Divide $-220$ by $-11$.

   **A)** $-22$    **D)** $-20$
   **B)** 20       **E)** 22
   **C)** 11

9. Multiply $(4 + 2)(-14 \times 5)$.

   **A)** $-510$   **D)** $-160$
   **B)** $-420$   **E)** $-320$
   **C)** 320

10. Find the value of $(4 + -2) \div -\frac{1}{2}$.

    **A)** $-2$    **D)** 4
    **B)** 16      **E)** $-4$
    **C)** 2

11. Multiply $2 \times -16 \times 6$.

    **A)** 172     **D)** 246
    **B)** 88      **E)** $-172$
    **C)** $-192$

12. Multiply $-14 \times 7 \times -\frac{1}{2}$.

    **A)** 49      **D)** 32
    **B)** $-56$   **E)** $-49$
    **C)** 63

13. Divide $-17$ by $-2$.

    **A)** $-8.5$  **D)** 8
    **B)** 8.5     **E)** $-9.5$
    **C)** $-8$

14. Divide $-288$ by $-9$.

    **A)** $-35$   **D)** 14
    **B)** $-26$   **E)** 32
    **C)** 26

**15.** Divide 114 by $-57$.

    **A)** 2            **D)** $-2$
    **B)** $-3$         **E)** 7
    **C)** $-7$

**16.** Find the sum of $-2$, 14, and $-75$.

    **A)** $-48$       **D)** $-83$
    **B)** $-99$       **E)** $-63$
    **C)** 14

**17.** Find the sum of $-89$, 27, and 14.

    **A)** 48          **D)** $-48$
    **B)** $-92$       **E)** 64
    **C)** 77

**18.** Evaluate $-77 - 28$.

    **A)** $-49$       **D)** 49
    **B)** 51          **E)** $-105$
    **C)** 105

**19.** Find the product of $(-12 \times 16) \times -4$.

    **A)** 288       **D)** $-394$
    **B)** 622       **E)** 979
    **C)** 768

**20.** Divide $-45$ by $-5$.

    **A)** 13          **D)** $-13$
    **B)** $-7$         **E)** 9
    **C)** 21

**21.** Find the value of $2 \times 2 \times 2 \times 2$.

    **A)** 12         **D)** 28
    **B)** 32         **E)** 22
    **C)** 16

**22.** Find $(45 \div 9) \times 16 + 2$.

    **A)** 120       **D)** 85
    **B)** 92         **E)** 78
    **C)** 75

**23.** Find the value of
$3 \times 16 \div (2 \times 2 \times 2)$.

    **A)** 10         **D)** 12
    **B)** 8          **E)** 6
    **C)** 63

**24.** Find $(121 \div 11) \times 3 + 5$.

    **A)** 27         **D)** 42
    **B)** 38         **E)** 65
    **C)** 77

# Algebraic Expressions

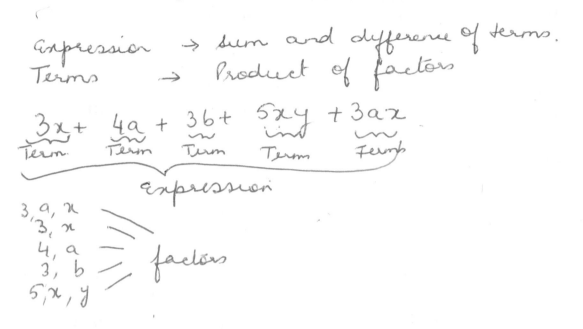

expression → sum and difference of terms.
Terms → Product of factors

$$\underset{\text{Term}}{3x} + \underset{\text{Term}}{4a} + \underset{\text{Term}}{3b} + \underset{\text{Term}}{5xy} + \underset{\text{Term}}{3az}$$

expression

3, a, x
3, x
4, a        }  factors
3, b
5, x, y

## 6.1 Translating Words into Mathematical Expressions and Formulas

Working with signed numbers is an introduction to using algebraic expressions. An **algebraic expression** is one in which letters or variables are used to represent numbers.

A letter or other type of symbol that represents a number is called a **literal number** or a **variable.**

If we know the numerical values of the variables and understand the arithmetic signs of an algebraic expression, then we can find the numerical value of any algebraic expression.

**Terms** are made up of numbers or variables combined by multiplication or division.

For example, $6DR$ is a term in which the factors 6, $D$ and $R$ are combined by multiplication; $M/4$ is a term in which the quantities $M$ and 4 are combined by division or in which the factors $M$ and $\frac{1}{4}$ are combined by multiplication.

An **expression** is a collection of terms combined by addition, subtraction, or both, and frequently grouped by parentheses, as in: $(3a + 2b)$, $(2c - 4c + 3b)$, $2x - 3y$.

To express word statements as formulas or as brief algebraic expressions, letters, and variables are substituted for words.

EXAMPLE 1: Express symbolically: What number increased by 6 gives 18 as a result?

SOLUTION: Substituting the letter $x$ for the unknown what number, we get
$$x + 6 = 18 \qquad x = ?$$

EXAMPLE 2: Express symbolically: The product of two numbers is 85. One is 5, find the other.

SOLUTION: $5x = 85$

$x = ?$

EXAMPLE 3: Express symbolically, 15 exceeds a certain number by 6. What is the number?

SOLUTION: $15 - 6 = x$

$x = ?$

EXAMPLE 4: Express symbolically: $\frac{2}{3}$ of a number is 20. Find the number.

SOLUTION: $\frac{2}{3}x = 20$

$x = ?$

In engineering, scientific, industrial, and commercial practice, it is common to express certain facts in algebraic **formulas.** The usual way is to state the formula with variables and to follow it immediately with an explanation (starting with the words *in which*) to make intelligible to the reader any variables that may require definition. Here are examples of this method of formula statement.

EXAMPLE 5: The cost equals the selling price minus the margin of profit.

SOLUTION: The formula is: $C = S - M$, in which $C$ stands for cost, $S$ for selling price, and $M$ for margin of profit.

EXAMPLE 6: The area of a rectangle equals the base times the height.

SOLUTION: The formula is: $A = bh$, in which $A$ stands for area, $b$ for base, and $h$ for height.

EXAMPLE 7: To determine the resistance in ohms of an electrical circuit, divide the number of volts by the number of amperes.

SOLUTION: The formula is: $O = V/A$, in which $O$ stands for ohms, $V$ for volts, and $A$ for amperes.

## Exercise Set 6.1

Write the following statements as equations.

1. The perimeter $(p)$ of a rectangle equals twice its length $(l)$ added to twice its width $(w)$.
2. The distance $(d)$ traveled by an object that moves at a given rate of speed $(r)$ for a given time $(t)$ equals the rate multiplied by the time.
3. To get the horsepower $(H)$ of an electric motor multiply the number of volts $(v)$ by the number of amperes $(a)$ and divide by 746.
4. Interest $(I)$ on money is figured by multiplying the principal $(P)$ by the rate $(R)$ by the time $(T)$.
5. The amperage $(A)$ of an electrical circuit is equal to the wattage $(W)$ divided by the voltage $(V)$.
6. Profit $(P)$ equals the margin $(M)$ minus the overhead $(O)$.
7. The distance $(d)$ that an object will fall in any given time $(t)$ is equal to the $t$ multiplied by itself time multiplied by 16.

8. The area ($A$) of a square figure is equal to the square of one of its sides ($S$).
9. Centigrade temperature ($C$) is equal to Fahrenheit temperature ($F$) minus $32°$, multiplied by $\frac{5}{9}$.
10. The speed ($R$) of a revolving wheel is proportional to the number of revolutions ($N$) it makes in a given time ($T$).

## 6.2 Adding and Subtracting Expressions

We can combine numbers that are represented by the same variables.

EXAMPLE 1: Add $5b$, $-11b$, and $14b$.

SOLUTION: As when combining signed numbers, we first combine expressions with the same signs.

$$5b + 14b = 19b$$

$$-11b = -11b$$

$$19b + (-11b) = 8b$$

We cannot add terms containing unlike symbols. For instance, if we let $b$ stand for books, and $p$ for plates, we know from arithmetic that we could not combine books and plates to get a single quantity of either. Therefore, to add quantities containing unlike symbols, collect **like terms** and express them separately in the answer.

EXAMPLE 2: Add $5b + 2p + 7p + 3b$.

SOLUTION: First, collect and simplify like terms.
$$5b + 3b = 8b$$

$$2p + 7p = 9p$$

Since we cannot add unlike terms, we write unlike terms separately. We get an algebraic expression containing two terms as the answer.

$$8b + 9p$$

If we are adding together several equations, we can stack them so that all like terms are placed in the same columns.

$$
\begin{array}{r}
3a - 4b + 2c \\
-8a + 6b - 3c \\
\underline{6a - 4b + 8c} \\
a - 2b + 7c
\end{array}
$$

To subtract algebraically, whenever we cannot directly subtract a smaller from a larger quantity of like sign, we mentally change the sign of the subtrahend and perform an addition.

$$
\begin{array}{r}
8a - 4b + 2c \\
\underline{5a - 6b + 8c} \\
3a + 2b - 6c
\end{array}
$$

$$8a - 5a = 3a; \ -4b + 6b = +2b;$$
$$2c - 8c = -6c.$$

EXAMPLE 3: Subtract $(5x + 2z)$ from $(2x - 11z)$.

SOLUTION: First, rewrite the subtraction. Then group like terms together and simplify.

$$(2x - 11z) - (5x + 2z) =$$

$$2x - 11z - 5x - 2z =$$

$$2x - 5x - 11z - 2z =$$

$$-3x - 13z =$$

**PRACTICALLY SPEAKING 6.2**

Eileen is balancing her checkbook for the month, on August 31. Here are her records for the month of August.

| NO. | ITEM | AMOUNT |
|---|---|---|
|  | Beginning Balance | $1,000 |
| 215 | Rent | $450 |
| 216 | Water Bill | $60 |
| Dep. |  | $600 |
| 217 | Shoes | $30 |
| 218 | Car Insurance | $200 |
| 219 | Groceries | $50 |

1. How much did Eileen spend on rent this month?
2. How much money does Eileen have left after totaling checks cashed and deposits?
3. What is Eileen's beginning balance for the month of September?

See Appendix F for the answers.

If we know the numerical values of the variables, we can simplify further.

EXAMPLE 4: Subtract 14x from 25x + 4, when x equals 3.

SOLUTION: First, set up the subtraction.

$$(25x + 4) - 14x = 11x + 4$$

Now substitute in the value for x.

$$11x + 4 = 11(3) + 4$$
$$= 33 + 4$$
$$= 37$$

**Exercise Set 6.2**

1. $-14d - 6d =$
2. $7b - 3b =$
3. $5x - 7x + 14x =$
4. $3a + 4b + 2a - 2b =$
5. $6a + 3b + 9a - 5 =$
6. $-22ab + 18ab =$
7. $(12x + 5t) - (14t - 6x) =$
8. $-71y + 14x - 15y - 13x =$
9. $19x + 12r - 66r + (-7x) =$
10. $56t - 38y + 16y - 90t =$
11. $678t - (-231t) =$
12. $45 - 51y + 30y - 11t =$
13. $190x - 24x + 26y - 67x =$

14. $25x - 18y + 65x - 112y =$

15. $392r - 89y + 72y =$

## 6.3 Exponents

**To square** a number is to use that number as a factor twice. Thus $4 \times 4$, or $4 \cdot 4 = 16$, and 16 is said to be the square of 4. This is also called raising a number to its **second power.** Using a number as a factor three times (for instance, $4 \times 4 \times 4 = 64$) is called raising it to the **third power.** The given case would be written $4^3$, and be read *four cubed;* $4^4$ is read *four to the fourth power,* $4^5$ is read *four to the fifth power,* etc.

A **power** of a number is the product obtained by multiplying the number by itself a given number of times. Raising a number to a given power is the opposite process of finding the corresponding root of a number.

To raise a given number to its indicated power, multiply the number by itself as many times as the power indicated. Thus, $3^5 = 3 \times 3 \times 3 \times 3 \times 3 = 243$. The small 5 used in writing $3^5$ is called an **exponent,** while the number 3 is called the **base.**

An **exponent** indicates the power to which a number is to be raised. Thus, $x^3$ means that $x$ is to be raised to the third power.

$a^2$ means $a \cdot a$. We read it as *a squared.* If $a = 3$, then $a^2 = 3 \cdot 3$ or 9.

$a^3$ means $a \cdot a \cdot a$. We read it as *a cubed.* If $a = 3$, then $a^3 = 3 \cdot 3 \cdot 3$ or 27.

$a^2 + b^3$ means that $b^3$ is to be added to $a^2$. If $a = 3$ and $b = 2$, then $a^2 + b^3 = 9 + 8$, or 17.

The small 2 and 3 placed to the right and slightly above the $a$ and the $b$ in writing $a^2$ and $b^3$ are called **exponents.**

The number $a$ is called the **base;** $a^2$ and $a^3$ are called **powers** of the base $a$.

$3a^2 - 2b^2$ means that $2b^2$ is subtracted from $3a^2$.

The $3a^2$ and $2b^2$ are known as **terms** in the algebraic expression.

The numbers placed before the letters are called **coefficients.** Thus, in $3a^2 - 2b^2$, 3 is called the coefficient of $a^2$, 2 is the coefficient of $b^2$. The coefficient so placed indicates multiplication, that is, $3a^2$ means $3 \times a^2$.

---

To raise a fraction to a given **power,** raise both the numerator and the denominator to the given power.

---

$$\left(\frac{1}{3}\right)^2 = \frac{1}{3} \times \frac{1}{3} = \frac{1}{9}$$

$$\left(\frac{3}{5}\right)^2 = \frac{3}{5} \times \frac{3}{5} = \frac{9}{25}$$

Any power or root of 1 is 1, because 1 multiplied or divided by 1 any number of times is 1.

Any number **without an exponent** is considered to be the first power or first root of itself. Neither the exponent nor the index 1 is written. Thus, $x$ means $x^1$.

Any number raised to the **zero power,** such as $5^0$, is equal to 1. The reason for this will appear when we consider the multiplication of powers of numbers.

When a number has a **negative exponent,** i.e., when the exponent is preceded by the minus sign, as in $3^{-3}$, it indicates the reciprocal of the indicated power of the number. Since $3^3 = 27$, $3^{-3} = \frac{1}{27}$, the reciprocal of 27. $12^{-2}$ means the reciprocal of $12^2$, or $\frac{1}{144}$.

## Powers of 10

$10^1 = 10$        $10^{-1} = \dfrac{1}{10}$ or 0.1

$10^2 = 10$        $10^{-2} = \dfrac{1}{100}$ or 0.01

$10^3 = 1,000$     $10^{-3} = \dfrac{1}{1,000}$ or 0.001

$10^4 = 10,000$    $10^{-4} = \dfrac{1}{10,000}$ or 0.0001

$10^5 = 100,000$   $10^{-5} = \dfrac{1}{100,000}$ or 0.00001

From this it is apparent that 10 raised to any positive power is equal to a multiple of 10 bearing as many zeros as are represented by the quantity of the exponent.

Also, 10 raised to any negative power is equal to a multiple of 10 containing as many decimal places as the quantity of the negative exponent.

The above forms are used for writing very large and very small numbers in an abbreviated way.

6,900,000 may be written as $6.9 \times 10^6$,

0.000008 may be written as $8 \times 10^{-6}$,

0.0000000235 may be written as
$$2.35 \times 10^{-8}$$

A positive exponent moves the decimal point a corresponding number of places to the right. Thus, $8.2 \times 10^7 = 82,000,000$.

A negative exponent moves the decimal point a corresponding number of places to the left. Thus, $6.3 \times 10^{-5} = 0.000063$.

Writing numbers as the product of a number between 1 and 10 times a power of 10 is known as **scientific notation.**

When we multiply variables together, our results may involve exponents.

EXAMPLE 1: What is the base in the expression $4^2$?

SOLUTION: The base is 4.

When we multiply together numbers like $2 \times 2$, we can represent the product by using exponents. Thus, $2 \times 2$ can be written $2^2$.

EXAMPLE 2: Write $2 \times 2 \times 2 \times 2$ using exponents.

SOLUTION: $2^4$

EXAMPLE 3: Write $\dfrac{1}{6} \times \dfrac{1}{6} \times \dfrac{1}{6}$ using exponents.

SOLUTION:

$$\frac{1}{6} \times \frac{1}{6} \times \frac{1}{6} =$$
$$\left(\frac{1}{6}\right)^3 =$$
$$6^{-3} =$$

## Exercise Set 6.3

In each of the following write the algebraic expression and find its numerical value if $x = 2$, $y = 3$, and $z = 4$.

1. $x$ added to $y$ =
2. $x$, $y$, and $z$ added together =
3. Twice $x$ added to twice $y$ =
4. $z$ subtracted from the sum of $x$ and $y$ =
5. The square of $x$ added to the square of $y$ =

**6.** 3 less than $y$ =

**7.** Twice the product of $x$ and $z$ =

Rewrite each expression using exponents.

**8.** $a \times a \times a \times a \times a$

**9.** $4 \times 4 \times 4 \times d \times d \times d$

**10.** $3 \times 3 \times 3 \times y \times y \times t \times t$

**11.** $7 \times 7 \times 7$

**12.** $3 \times 3 \times b \times b$

Perform the indicated operations.

| | |
|---|---|
| **13.** $6^2 =$ | **18.** $43 \times 10^6 =$ |
| **14.** $9^3 =$ | **19.** $6.2 \times 10^5 =$ |
| **15.** $4^{-3} =$ | **20.** $2.8 \times 10^{-7} =$ |
| **16.** $432^2 =$ | **21.** $25 \times 10^{-4} =$ |
| **17.** $8^5 =$ | **22.** $12.2 \times 10^7 =$ |

## 6.4 Multiplying and Dividing Expressions

Multiplication can be indicated in four ways in algebra; $a$ multiplied by $b$ can be written $a \times b$, $a \cdot b$, $(a)(b)$, or $ab$. That is, multiplication can be expressed by a cross $\times$, by a dot $\cdot$, by adjacent parentheses, and by directly joining a letter and its multiplier with no sign between them. Thus $2a$ means 2 times $a$, and $ab$ means $a$ times $b$.

### Laws of Exponents

To multiply powers of the same base, add their exponents.

$$2^2 \text{ times } 2^3 = 2^5$$

PROOF: $2^2 = 4$; $2^3 = 8$; $2^5 = 32$

$$4 \times 8 = 32 \cdot$$

To divide powers of the same base, subtract the exponent of the divisor from the exponent of the dividend.

$$3^5 \div 3^3 = 3^2$$

PROOF: $3^5 = 243$; $3^3 = 27$; $3^2 = 9$

$$243 \div 27 = 9$$

It will now become apparent why *any* number with an exponent of zero is equal to 1. According to the laws just stated—

$$x^3 \times x^0 = x^{3-0} = x^3$$

because if equals are multiplied by equals the products are equal;

$$\text{but } x^3 \times 1 = x^3$$

$$\text{so } x^0 = 1$$

To generalize this fact, let $n$ denote any positive exponent whatever. Then $x^n \times x^0 = x^n$ and $x^0$ necessarily equals 1. The same conclusion will be reached if the process is division and the exponents are subtracted.

$$x^n \div x^0 = x^{n-0} = x^n, \therefore x^0 = 1$$

EXAMPLE 1: Multiply $4x$ by $(12x + 4y)$.

SOLUTION: Multiply $4x$ by each term inside the parenthesis.

$$4x (12x + 4y) = 4x (12x) + 4x(4y)$$
$$= 48x^2 + 16xy$$

Note that we can stack expressions and multiply them. Multiplication is performed like this:

$$
\begin{array}{r}
a^2 - 2ab + b^2 \\
a - b \\
\hline
a^3 - 2a^2b + \phantom{2}ab^2 \\
-\phantom{a^3} a^2b + 2ab^2 - b^3 \\
\hline
a^3 - 3a^2b + 3ab^2 - b^3
\end{array}
$$

Each term in the multiplicand is multiplied separately by $a$ and then by $b$. Like terms are set under each other and the whole is added. $+ \times +$ gives $+$; $- \times -$ gives $+$; $+ \times -$ gives $-$.

EXAMPLE 2: Divide $25x$ by $5x$.

SOLUTION: Note what happens to the variable when we divide.

$$25x \div 5x = 5$$

Note that we can also stack expressions to divide. Division is performed like this:

$$\frac{3a^2b + 3ab^2 + 3a}{3a} = ab + b^2 + 1,$$

$$\text{or } b^2 + ab + 1$$

$3a$ is a factor of each term in the dividend. Separate divisions give us $ab + b^2 + 1$. This is changed to $b^2 + ab + 1$ because it is customary to place algebraic terms in the order of their highest powers.

To reduce a fractional expression to its lowest terms, resolve the numerator and the denominator into their prime factors and cancel all the common factors, or divide the numerator and the denominator by their highest common factor.

EXAMPLE 3: Reduce $\dfrac{12a^2b^3c^4}{9a^3bc^2}$.

SOLUTION:

$$\frac{12a^2b^3c^4}{9a^3bc^2} = \frac{2 \times 2 \times 3a^2bc^2(b^2c^2)}{3 \times 3a^2bc^2(a)}$$

$$= \frac{4b^2c^2}{3a}$$

The numerical parts of the fraction are separated into their prime factors, and the algebraic parts are divided by their highest common factor. The terms that cancel out are then eliminated. As a guide for determining the highest common factor, note that such a factor is made up of the lower (or lowest) of the given powers of each letter involved.

To reduce a mixed expression to a fraction, multiply the integer part of the expression by the denominator of the fraction; add to this product the numerator of the fraction and write under this result the given denominator.

EXAMPLE 4: Reduce $x + 1 + \dfrac{x + 1}{x - 1}$ to a fraction.

SOLUTION:

$$(x + 1)\left(\frac{x - 1}{x - 1}\right) + \frac{x - 1}{x - 1}$$

$$= \frac{x^2 - 1 + x + 1}{x - 1} = \frac{x^2 + x}{x - 1}$$

To reduce fractions to their lowest common denominator, find the lowest common multiple of the denominators and proceed on the same principles that govern arithmetical fractions.

The product of two or more fractions is equal to the product of the numerators multiplied together, divided by the product of the denominators multiplied together.

EXAMPLE 5: Multiply $\dfrac{7x}{5y}$ by $\dfrac{3a}{4c}$.

SOLUTION: $\dfrac{7x}{5y} \cdot \dfrac{3a}{4c} = \dfrac{21ax}{20cy}$

EXAMPLE 6: Multiply $\dfrac{2x}{x - y}$ by $\dfrac{x^2 - y^2}{3}$.

SOLUTION: $\left(\dfrac{2x}{x - y}\right)\left(\dfrac{x^2 - y^2}{3}\right)$

$= \dfrac{2x(x + y)(x - y)}{3(x - y)}$

$= \dfrac{2x(x + y)}{3}$

EXAMPLE 7: Simplify $\dfrac{2a - 4b}{4} - \dfrac{a - b + c}{3}$

$+ \dfrac{a - b + 2c}{12}$.

SOLUTION: $\dfrac{2a - 4b}{4} - \dfrac{a - b + c}{3}$

$+ \dfrac{a - b + 2c}{12}$

$= \dfrac{\begin{array}{c}6a - 12b - 4a + 4b \\ - 4c + a - b - 2c\end{array}}{12}$

$= \dfrac{3a - 9b - 6c}{12}$

$= \dfrac{a - 3b - 2c}{4}$

EXAMPLE 8: Simplify $\dfrac{a + 2x}{a - 2x} - \dfrac{a - 2x}{a + 2x}$.

SOLUTION: $\dfrac{a + 2x}{a - 2x} - \dfrac{a - 2x}{a + 2x}$

$= \dfrac{(a + 2x)^2 - (a - 2x)^2}{a^2 - 4x^2}$

$= \dfrac{\begin{array}{c}a^2 + 4ax + 4x^2 \\ - a^2 + 4ax - 4x^2\end{array}}{a^2 - 4x^2}$

$= \dfrac{8ax}{a^2 - 4x^2}$

## Exercise Set 6.4

Simplify.

1. $16x \times 11xy + 14y =$
2. $(10x + 4y) + 23x - 75y =$
3. $66y \div 3 - 34y =$
4. $45x \div 9x + (23x - 14) =$
5. $18xy \div 2y =$
6. $11x \times 13y - 5x =$
7. $12r \times 65t + (34t - 11t) =$
8. $58 \times 19w - 45 \div 5 =$
9. $45x \div 6t \times 11x =$
10. $33x \div 11x =$
11. Reduce $\dfrac{45x^3y^3z}{36abx^2y^2z}$ to its lowest terms.
12. Reduce $a + \dfrac{ax}{a - x}$ to a fraction.
13. Reduce $1 + \dfrac{c}{x - y}$ to a fraction.
14. Reduce $\dfrac{x + a}{b}$, $\dfrac{a}{b}$, and $\dfrac{a - x}{a}$ to fractions with the LCD.
15. Multiply $\dfrac{x^2 - 4}{3}$ by $\dfrac{4x}{x + 2}$.

**16.** Divide $\dfrac{3x}{2x - 2}$ by $\dfrac{2x}{x - 1}$.

**17.** Add $\dfrac{x + y}{2}$ and $\dfrac{x - y}{2}$.

**18.** Add $\dfrac{2}{(x - 1)^3}$, $\dfrac{3}{(x - 1)^2}$ and $\dfrac{4}{x - 1}$.

**19.** Subtract $2a - \dfrac{a - 3b}{c}$ from $4a + \dfrac{2a}{c}$.

**20.** Subtract $\dfrac{x}{a + x}$ from $\dfrac{a}{a - x}$.

## 6.5 Equations

A **formula** is a method of expressing a rule by the use of symbols or letters.

At the same time it must be remembered that a formula is an equation. And what is an equation?

An **equation** is a statement that two expressions are equal.

For example, $D = BR$ states that $D$, the discount, is equal to $B$, the base, multiplied by $R$, the rate of discount. Before we can start working with formulas and equations there are a few things that have to be learned about them.

An equation has two equal sides or members. In the equation $D = BR$, $D$ is the left side and $BR$ is the right side.

When we solve equations, we find the value of the unknown or literal number in relation to other numbers in the equation. To do this we must learn the following rules of procedure for treating equations. Primarily, what we do to one side of an equation, we must also do to the other.

---

The same number may be added to both sides of an equation without changing its equality.

---

EXAMPLE 1: If $x - 4 = 6$, what does $x$ equal?

SOLUTION: Add 4 to both sides.

$$x - 4 + 4 = 6 + 4.$$
$$x = 10$$

To check the solution of algebraic examples, substitute the value of the unknown quantity as determined in the answer for the corresponding symbol in the original equation. If both sides produce the same answer, the answer is correct.

EXAMPLE 2: Check the correctness of 10 as the solution of $x - 4 = 6$.

SOLUTION:

$x - 4 = 6$   Original equation.

$10 - 4 = 6$   Substitute the answer for the symbol.

$6 = 6$   Proof of correctness.

---

The same number may be subtracted from both sides of an equation.

---

EXAMPLE 3: If $n + 6 = 18$, what does $n$ equal?

SOLUTION: Subtract 6 from both sides of the equation.

$$n + 6 - 6 = 18 - 6$$
$$n = 12$$

Check by substituting 12 for $n$ in the original equation. Thus, $n + 6 = 18$ becomes $12 + 6 = 18$ or $18 = 18$, which is correct.

---

Both sides of an equation may be multiplied by the same number.

---

EXAMPLE 4: If ⅓ of a number is 10, find the number.

SOLUTION: Multiply both sides by 3; then cancel.

$$\frac{1}{3}n \text{ or } \frac{n}{3} = 10$$

$$\frac{n}{3} \times 3 = 10 \times 3$$

$$\frac{n}{\cancel{3}} \times \cancel{3} = 10 \times 3$$

$$n = 30$$

Check the answer.

---

Both sides of an equation may be divided by the same number.

---

EXAMPLE 5: Two times a number is 30. What is the number?

SOLUTION: Divide both sides by 2.

$$2n = 30$$

$$\frac{2n}{2} = \frac{30}{2}$$

$$n = 15$$

Check the answer.

**Transposition** is the process of moving a quantity from one side of an equation to the other side by changing its sign of operation. This is exactly what has been done in carrying out the rules in the four examples above.

Division is the operation opposite to multiplication.

Addition is the operation opposite to subtraction.

Transposition is performed in order to obtain an equation in which the unknown quantity is on one side and the known quantity is on the other.

---

A term may be transposed from one side of an equation to the other if its sign is changed from + to −, or from − to +.

---

A factor, or multiplier, may be removed from one side of an equation by making it a divisor in the other. A divisor may be removed from one side of an equation by making it a factor in the other.

---

Observe again the solution to Example 2.

$$x - 4 = 6$$
$$x = 6 + 4$$
$$x = 10$$

To get $x$ by itself on one side of the equation, the $-4$ was transposed from the left to the right side and made $+4$.

Observe again the solution to Example 3.

$$n + 6 = 18$$

$$n = 18 - 6$$

$$n = 12$$

To get $n$ by itself on one side of the equation, the $+6$ was transposed from the left to the right side and made $-6$.

Observe again the solution to Example 4.

$$\frac{n}{3} = 10$$

$$n = 10 \times 3$$

$$n = 30$$

To get $n$ by itself on one side of the equation, the divisor 3 on the left was changed to to the multiplier $3(\frac{3}{1})$ on the right.

Observe again the solution to Example 5.

$$2n = 30$$

$$n = \frac{30}{2}$$

$$n = 15$$

To get $n$ by itself on one side of the equation, the multiplier 2 on the left was changed to the divisor 2 on the right.

Note that transposition is essentially a shortened method for performing like op-erations of addition, subtraction, multiplication, or division on both sides of the equation.

Changing $x - 4 = 6$ to $x = 6 + 4$ is the same as adding 4 to both sides:

$$\begin{array}{rcl} x - 4 &=& 6 \\ +4 &=& +4 \\ \hline x &=& 10 \end{array}$$

Changing $n + 6 = 18$ to $n = 18 - 6$ is the same as subtracting 6 from both sides:

$$\begin{array}{rcl} n + 6 &=& 18 \\ -6 &=& -6 \\ \hline n &=& 12 \end{array}$$

Changing $\frac{n}{3} = 10$ to $n = 10 \times 3$ is the same as multiplying both sides by 3:

$$\frac{n}{3} \times 3 = 10 \times 3,$$

in which the 3's on the left cancel.

Changing $2n = 30$ to $n = \frac{30}{2}$ is the same as dividing both sides by 2:

$$\frac{2n}{2} = \frac{30}{2},$$

in which the 2's on the left cancel.

When terms involving the unknown quantity occur on both sides of the equation, perform such transpositions as may be necessary to collect all the unknown terms on one side (usually the left) and all the known terms on the other.

EXAMPLE 6: If $3x - 6 = x + 8$ what does $x$ equal?

SOLUTION:

$3x = x + 8 + 6$     Transpose $-6$ from left to right.

$3x - x = 14$     Transpose $x$ from right to left.

$2x = 14$     Transpose 2 as a multiplier from left to a divisor at the right.

$x = \dfrac{14}{2}$

$x = 7$

Check:

$3x - 6 = x + 8$

$21 - 6 = 7 + 8$     Substitute 7 for $x$ for proof of correctness.

$15 = 15$

When using an algebraic formula, it may be necessary to change its form. Such changes are effected by transposition.

EXAMPLE 7: If $R = \dfrac{WC}{L}$, solve for $W$, $C$, and $L$.

SOLUTION:

$R = \dfrac{WC}{L}$     Original formula.

$\dfrac{LR}{C} = W$     To separate $W$, $C$ and $L$ are transposed.

$\dfrac{LR}{W} = C$     To separate $C$, $L$ and $W$ are transposed.

$L = \dfrac{WC}{R}$     To separate $L$, $L$ and $R$ are transposed.

## Exercise Set 6.5

Solve transposition.

1. $p + 3 = 8$     $p = ?$
2. $2n = 25$     $n = ?$
3. $\frac{1}{2}x = 14$     $x = ?$
4. $5c - 3 = 27$     $c = ?$
5. $18 = 5y - 2$     $y = ?$
6. $\frac{2}{3}n = 24$     $n = ?$

---

**PRACTICALLY SPEAKING 6.5**

Jenny goes to Talbot's to buy a dress. Since there is a 35%-off sale on some merchandise, she thinks she may be able to find a bargain. Jenny finds two dresses that she would like to buy. One costs $150, but is marked 35% off, so the dress will cost $150 minus 35%. The second dress costs $115, but there is no discount.

1. Which dress will be less expensive to buy?
See Appendix F for the answer.

7. $\dfrac{a}{2} + \dfrac{a}{4} = 36$    $a = ?$

8. $W = \dfrac{b}{c}$    $b = ?$

9. $V = \dfrac{W}{A}$    $A = ?$

10. $H = \dfrac{P}{AW}$    $W = ?$

## Chapter 6 Glossary

**Algebraic Expression** An expression where letters are used to represent numbers.

**Base** The number used as the factor when finding a power.

**Coefficient** The numerical factors in a term.

**Equation** A statement that two expressions are equal.

**Exponent** The number that indicates how many times the base is to be used as a factor.

**Expression** A collection of terms combined by addition, subtraction, or both.

**Formula** A mathematical relationship stated using variables.

**Like Terms** Terms that have exactly the same variables raised to exactly the same powers.

**Literal Number or Variable** A letter or other type of symbol that represents a number.

**Negative Exponent** When the exponent is preceded by a minus sign, it indicates that the number is the reciprocal of the indicated power of the number.

**Power** A power of a number is the product obtained by multiplying the number by itself a given number of times.

**Scientific Notation** A method of writing a number as a product, so that it appears as a number between 1 and 10 multiplied by a power of 10.

**Terms** Numbers or variables combined by multiplication or division.

**Transposition** The process of moving a quantity from one side of an equation to the other side by changing its sign of operation.

## Chapter 6 Test

For each problem, five answers are given. Only one answer is correct. After you solve each problem, check the answer that agrees with your solution.

Simplify each of the following expressions.

1. $-4 \times -16$.

   A) 25    D) $-72$
   B) $-64$    E) 72
   C) 64

2. $-33 \times 11$.

   A) 363    D) $-333$
   B) $-363$    E) $-330$
   C) 330

3. $-54 \times -13$.

   A) 512    D) 702
   B) $-512$    E) 744
   C) $-702$

4. $221 \div 13$.

   A) 12    D) 17
   B) 15    E) $-17$
   C) $-12$

**5.** 136 ÷ 8.

   **A)** 17        **D)** 19
   **B)** −17     **E)** −21
   **C)** 13

**6.** −6 + (−11) + 14.

   **A)** −31     **D)** −3
   **B)** −13     **E)** 13
   **C)** 3

**7.** −46 − (−11) + 14.

   **A)** 21       **D)** −31
   **B)** 17       **E)** −17
   **C)** −21

**8.** 231 − (−75) + (−140).

   **A)** −520    **D)** 166
   **B)** 406     **E)** 209
   **C)** −120

**9.** 716 − (−257) + 146.

   **A)** 579     **D)** 1,267
   **B)** −17     **E)** 1,109
   **C)** 857

**10.** −540 ÷ 12.

   **A)** −35     **D)** 40
   **B)** 15      **E)** −45
   **C)** 75

**11.** 364 ÷ −70.

   **A)** −32     **D)** −91
   **B)** 47      **E)** 91
   **C)** −52

**12.** 289 ÷ 17.

   **A)** 13      **D)** 17
   **B)** 11      **E)** −13
   **C)** −11

**13.** −13 × 45 ÷ −9.

   **A)** 260     **D)** 65
   **B)** 165     **E)** 95
   **C)** −165

Translate each written expression into a mathematical expression.

**14.** Twelve minus fourteen.

   **A)** 4 − 2     **D)** 6
   **B)** 12 − 14   **E)** 12 − 4
   **C)** 14 − 12

**15.** Multiply two hundred ten by fourteen.

   **A)** 210 ÷ 14       **D)** 210 × 14
   **B)** 200 × 10 ÷ 14  **E)** −210 ÷ 14
   **C)** 200 × 14

**16.** Add sixteen to the product of eleven and seventy.

   **A)** (11 × 70) + 16  **D)** 16 + 70 ÷ 11
   **B)** 77 + 16        **E)** (11 × 70) ÷ 16
   **C)** 16 + 11 × 70

**17.** Subtract six hundred from eleven hundred twenty.

   **A)** 110 + 20 − 600 **D)** 1,120 − 600
   **B)** 1,100 − 600    **E)** 1,020 − 600
   **C)** 112 + 600

Solve each equation for $x$.

**18.** $2x - 4 = 20$. ✓

   **A)** $x = 16$     **D)** $x = -14$
   **B)** $x = -8$     **E)** $x = 8$
   **C)** $x = 12$

**19.** $6x + 24 = 3x$. ✓

   **A)** $x = 8$      **D)** $x = -24$
   **B)** $x = -8$     **E)** $x = 18$
   **C)** $x = 24$

**20.** $5x - 13 = 2x + 8$. ✓

   **A)** $x = 7$      **D)** $x = -7$
   **B)** $x = -5$     **E)** $x = 4$
   **C)** $x = 5$

# Polynomials

## 7.1 Roots

For the number 9, 3 and 3 are equal factors; and for 8, 2, 2, and 2 are equal factors. These equal factors are called **roots** of the number. Thus:

The number 3 is a root of 9.

The number 2 is a root of 8.

A **root** of a number is therefore one of the equal factors which, if multiplied together, produce the number.

The **square root** of a number is one of **two** equal factors which, if multiplied together, produce that number.

$3 \times 3 = 9$, therefore 3 is the square root of 9.

The **cube root** of a number is one of **three** equal factors which if multiplied together produce that number.

$3 \times 3 \times 3 = 27$, therefore 3 is the cube root of 27.

A **fourth root** of a number is one of four equal factors; the fifth root is one of five, and so on.

The square root is the one most frequently used in mathematics.

The sign indicating square root is $\sqrt{\phantom{x}}$. It is placed over the number whose root is to be found. $\sqrt{25}$ means the square root of 25. It is called the **square root** sign or **radical sign.**

To indicate a root other than square root a small figure called the **index** of the root is placed in the radical sign. Thus: $\sqrt[3]{8}$ means the cube root of 8.

The square root of 4 = 2, of 36 = 6, of 49 = 7.

To check that we have obtained the correct square root of a number, multiply it by itself. If the product is equal to the original number, the answer is correct.

Not all numbers have exact square roots. Nor can we always determine square root by inspection as we have done above. (Inspection means "trial and error.") There is an arithmetic method of extracting the square root of a number whereby an answer may be found that will be correct to any necessary or desired number of decimal places.

## Method for Finding Square Roots

To find the square root of 412,164.

**1.** Place the square root sign over the number, and then, beginning at the right, divide it into periods or groups, of two figures each. Connect the digits in each period with tie-marks as shown. In the answer there will be one digit for each period.

$$\sqrt{41\ 21\ 64}$$

**2.** Find the largest number which, when squared, is contained in the first left-hand period. In this case 6 is the number. Write 6 in the answer over the first period. Square it, making 36, and subtract 36 from the first period. Bring down the next period, making the new dividend 5 21.

$$\begin{array}{r} 6\phantom{aaaaa} \\ \sqrt{41\ 21\ 64} \\ \underline{36}\phantom{aaaa} \\ 5\ 21\phantom{aa} \end{array}$$

**3.** Multiply the root 6 by 2, getting 12. Place the 12 to the left of 5 21, since 12 is the new trial divisor. Allow, however, for one more digit to follow 12. The place of this missing digit may be indicated by a question mark. To find the number belonging in this place, ignore (cover over) the last number in the dividend 5 21, and see how many times 12 goes into 52. Approximately 4. Place the 4 above its period, 21, and put it in place of the ? in the divisor.

$$\begin{array}{r} 6\quad 4\phantom{aaa} \\ \sqrt{41\ 21\ 64} \\ \underline{36}\phantom{aaaa} \\ 12^2_4\ |\ \overline{5\ 21}\phantom{aa} \end{array}$$

**4.** Multiply the divisor 124 by the new number in the root, 4.124 × 4 = 496. Place this product under 5 21 and subtract. Bring down the next period, 64.

$$\begin{array}{r} 6\quad 4\phantom{aaa} \\ \sqrt{41\ 21\ 64} \\ \underline{36}\phantom{aaaa} \\ 124\ |\ \overline{5\ 21}\phantom{aa} \\ \underline{4\ 96}\phantom{aa} \\ 25\ 64 \end{array}$$

**5.** Multiply 64 by 2 to get 128 as the new trial divisor. 128 goes into 256 two times. Place the 2 above the next period in the root and also in the divisor. Then multiply the divisor 1282 by the new root 2, to get 25 64. Subtracting, the remainder is zero; 642 is therefore the exact square root.

$$\begin{array}{r} 6\quad 4\quad 2\phantom{a} \\ \sqrt{41\ 21\ 64} \\ \underline{36}\phantom{aaaaa} \\ 124\ |\ \overline{5\ 21}\phantom{aaa} \\ \underline{4\ 96}\phantom{aaa} \\ 128^2_2\ |\ \overline{25\ 64} \\ \underline{25\ 64} \\ 0 \end{array}$$

**6.** CHECK: 642 × 642 = 412,164.

## Finding the Square Root of Decimals

A slight variation in method is necessary when it is required to find the square root of a decimal figure.

Mark off periods beginning at the decimal point. Count to the right for the decimal quantities and to the left for the whole numbers. If the last period of the whole numbers contains one figure, leave it by itself, but remember that in such a case the first figure in the root cannot be more than 3 because the square of any number greater than 3 is a two-place number. If the last period of the decimal numbers contains only one figure, you may add a zero to it. This is because two digits are necessary to make up a period, while the addition of a zero at the right of a decimal figure does not change its value.

The square root of a decimal will contain as many decimal places as there are periods, or half as many decimal places as the given number.

The operations in obtaining the square root of a decimal number are the same as for whole numbers.

Follow the steps in the example following.

EXAMPLE 1: Find the square root of 339.2964.

SOLUTION:

**1.** Beginning at the decimal point, mark off periods to the left and right.

**2.** 1 is the largest whole-number square root that is contained in 3, which constitutes the first period.

**3.** Place a decimal point in the root after the 8 because the root of the next period has a decimal value.

**4.** Bring down 29 next to the 15, making 1529 the new dividend. Multiply the root 18 by 2, making 36 the new divisor.

$$\begin{array}{r} 1\ \ 8.\ 4\ \ 2 \\ \sqrt{3\ \ 39.29\ 64} \\ \underline{1} \\ 2_8^?\overline{\big|2\ 39} \\ \underline{2\ 24} \\ 36_4^?\overline{\big|\ \ 15\ 29} \\ \underline{14\ 56} \\ 368_2^?\overline{\big|\ \ \ \ 73\ 64} \\ \underline{73\ 64} \\ 0 \end{array}$$

**5.** Covering the 9 of 1529, 36 seems to be contained about 4 times in this number. Place a 4 in the root above 29, and multiply 364 by 4 to get 1456. Subtract this from 1529.

**6.** Bring down the 64 and repeat the previous process. Since the number is a perfect square, the remainder is zero.

When the given number is not a perfect square, add zeros after the decimal point, or after the last figure if the original number is already in decimal form, and carry out the answer to the required or desired number of decimal places. Usually two places are sufficient.

In working a square root example, when a divisor is larger than the corresponding dividend, write zero in the trial divisor and bring down the next period.

EXAMPLE 2: Find the square root of 25.63 to three decimal places.

SOLUTION:

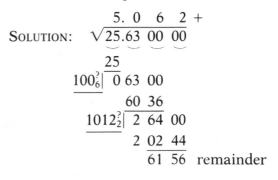

$$\begin{array}{r} 5.\ \ 0\ \ \ 6\ \ \ 2\ + \\ \sqrt{25.63\ 00\ 00} \\ \underline{25} \\ 100_6^?\overline{\big|\ 0\ 63\ 00} \\ \underline{60\ 36} \\ 1012_2^?\overline{\big|\ 2\ 64\ 00} \\ \underline{2\ 02\ 44} \\ 61\ 56\ \ \text{remainder} \end{array}$$

To find the square root of a fraction, determine separately the square roots of the numerator and of the denominator, and reduce to lowest terms or to a decimal.

EXAMPLE 3: Find $\sqrt{\dfrac{33}{67}}$.

SOLUTION: $\sqrt{\dfrac{33}{67}} = \dfrac{\sqrt{33}}{\sqrt{67}} = \dfrac{5.745}{8.185} = .701$

## Use of Square Roots

In studying the following example, read step by step the rule that follows it and note how the example illustrates the rule.

EXAMPLE 4: What is the cube root of 264,609,288?

SOLUTION:

$$\sqrt[3]{264\ 609\ 288} = 6\ 4\ 2$$

$$6^3 = 216$$

First Part. Div.
$3 \times 60^2 = 10{,}800$ | 48 609
$3 \times 60 \times 4 = 720$
$4^2 = 16$
First Comp. Div. $11{,}536$ | 46 144
2 465 288

Second Part. Div.
$3 \times 640^2 = 1{,}228{,}800$
$3 \times 640 \times 2 = 3{,}840$
$2^2 = 4$
Second Comp. Div.
$1{,}232{,}644$ | 2 465 288

The following rule is more readily understood if we bear in mind the formula for the cube of the sum of two numbers.

$$(a + b)^3 = a^3 + 3a^2b + 3ab^2 + b^3$$

**Rule: 1.** Separate the given number into periods of three figures each, beginning at the right, and place over it the radical sign with the proper index.

The extreme left-hand period may contain one, two, or three figures.

**2.** Determine the greatest cube that is smaller than the first left-hand period, and write its cube root, in the position shown, as the first figure of the required root.

This root corresponds to $a$ in the formula.

**3.** Subtract the cube of this root from the first period and bring down the next period.

**4.** Multiply this root mentally by 10 and write three times the square of this as a partial divisor.

**5.** Make a trial division to determine what the next figure in the root will be and write it in its proper place.

**6.** Add to the partial divisor (1) the product of three times the first part of the root considered as tens multiplied by the second part of the root; and (2) the square of the second part of the root. The sum of these numbers is the complete divisor.

**7.** Multiply the complete divisor by the second part of the root and subtract the product from the new dividend.

**8.** Bring down the next period and continue the same process until all the figures of the root have been determined.

## Approximate Roots of Fractions

We have seen that the square root of a fraction is the square root of its numerator placed over the square root of its denominator, subject to further reduction or to conversion to a decimal.

When the terms of a fraction are not perfect squares it is often desirable to approximate a square root without going to the trouble of making an exact calculation. This is done by multiplying the terms of the fraction by any number that will make the denominator a perfect square, as in the following example.

EXAMPLE 5: What is the approximate square root of $19/8$?

SOLUTION: $19/8 = 38/16$, of which the approximate square root, $6/4$, is correct to within $1/4$; or $19/8 \times 32/32 = 608/256$, of which the approximate square root, $25/16$, is correct to within $1/16$

We select a factor that will make the denominator a perfect square. We then extract the square root of the denominator and the square root of the perfect square that is nearest to the numerator. If we write the fraction as $38/16$, the square root of the denominator is 4 and the square root of the nearest perfect square to 38 is 6. The resulting approximate square root, $6/4$, reducible to $3/2$, is correct to within $1/4$.

If we want a closer approximation than this, we multiply by a larger factor. Using 32 as a factor, we get $608/256$. The square root of the denominator is 16. The nearest perfect square to 608 is 625, the square root of which is 25. The resulting approximate square root, $25/16$, is correct to within $1/16$.

It will be noted that the larger the factor the more closely will the result approximate the correct value.

The approximate cube root of a fraction may be found by a similar process.

EXAMPLE 6: Find the approximate cube root of $173/32$.

SOLUTION: $173/32 = 346/64$, of which the approximate cube root, $7/4$, is correct to within $1/4$; or $173/32 = 2,768/512$, of which the approximate cube root, $14/8$, is correct to within $1/8$

The denominator has been multiplied by two different factors in order to demonstrate again that the higher factor produces the more nearly accurate answer. It will be noted that the final result in both cases has the same ultimate value since $14/8 = 7/4$. If, however, we had not worked out the second solution we would not know that $7/4$ is actually correct to within $1/8$.

## Higher Roots

If the index of a higher root contains no other prime factors than 2 and 3, we can find the required root by repeated extraction of square or cube roots, according to the nature of the problem.

EXAMPLE 7: What is the fourth root of 923521?

SOLUTION: $\sqrt{923521} = 961$
$\sqrt{961} = 31$

Since the fourth power of a number is its square multiplied by its square, we find the fourth root of a given number representing such a power by extracting the square root of the square root.

EXAMPLE 8: What is the sixth root of 191102976?

SOLUTION: $\sqrt{191102976} = 13824$
$$\sqrt[3]{13824} = 24$$

The sixth root is found by taking the cube root of the square root. The order of making the extractions is of course immaterial.

Higher roots with indexes that are prime to 2 and 3 can be found by methods based on the same general theory as that underlying the methods for extracting square and cube roots. In actual practice, however, higher roots are more commonly found by the use of logarithms. Some scientific calculators have keys for calculating higher roots.

When a number has a **fractional exponent** with a numerator of 1, as has $x^{1/2}$, it signifies that the corresponding root is to be taken of the number. In other words, $16^{1/2} = \sqrt{16} = 4$.

When a **fractional exponent** has a numerator greater than one, as has $x^{4/5}$, the numerator indicates the power to which the number is to be raised, while the denominator indicates the root that is to be taken. Accordingly, $4^{3/2} = \sqrt{4^3} = \sqrt{64} = 8$. To reverse this example, $8^{2/3} = \sqrt[3]{8^2} = \sqrt[3]{64} = 4$.

## Exercise Set 7.1

Find the roots indicated and then check.

1. $\sqrt{64}$
2. $\sqrt{100}$
3. $\sqrt{81}$
4. $\sqrt[3]{27}$
5. $\sqrt[3]{125}$
6. $\sqrt{144}$
7. $\sqrt[3]{1,000}$
8. $\sqrt{1}$

9. $\sqrt{.04} = \sqrt{.2 \times .2} = .2$
10. $\sqrt{.09}$
11. $\sqrt{5,329}$
12. $\sqrt{1,225}$
13. $\sqrt{2,937.64}$
14. $\sqrt{312,649}$
15. $\sqrt{428}$ to two places
16. $\sqrt{676}$
17. $\sqrt{1,849}$
18. $\sqrt{3,136}$
19. $\sqrt{7,225}$
20. $\sqrt{9,409}$

Find the required roots (approximate in the case of fractions).

21. $\sqrt[3]{2,460,375}$
22. $\sqrt[3]{11,089,567}$
23. $\sqrt[3]{40,353,607}$
24. $\sqrt[3]{403,583,419}$
25. $\sqrt[3]{115,501,303}$
26. $\sqrt{2/3} \, (\times \, 48/48)$
27. $\sqrt{38/5} \, (\times \, 5/5)$
28. $\sqrt{45/7} \, (\times \, 343/343)$
29. $\sqrt{10\frac{1}{2}} \, (\times \, 200/200)$
30. $\sqrt{7\frac{1}{8}} \, (\times \, 20,000/20,000)$
31. $\sqrt[3]{2/3} \, (\times \, 72/72)$
32. $\sqrt[3]{2/3} \, (\times \, 15,552/15,552)$
33. $\sqrt[3]{2/3} \, (\times \, 124,416/124,416)$
34. $\sqrt[3]{5^{13}/32} \, (\times \, 2/2)$
35. $\sqrt[3]{125/256} \, (\times \, 2/2)$
36. $\sqrt[4]{6,561}$
37. $\sqrt[6]{117,649}$
38. $\sqrt[4]{29^{52}/81}$
39. $\sqrt[4]{104^{536}/625}$
40. $\sqrt[6]{11^{25}/64}$

## 7.2 Factoring Binomials

A **factor** of a number is an exact divisor of that number. Thus 2 is a factor of 6 because $6 \div 2 = 3$ exactly; 3 is the other factor of 6.

A **monomial** is an algebraic expression of one term. Thus, $8a$ and $16a^2b$ and $\sqrt{3ax}$ are monomials.

A **polynomial** is an algebraic expression of more than one term. Thus, $a + b$, and $a^2 + 2ab + b^2$, and $a^3 + 3a^2b + 3ab^2 + b^3$, are three different polynomials.

A **binomial** is a polynomial that contains two terms. Thus, $a + b$, and $a + 1$, and $\sqrt{2} + \sqrt{3}$, are binomials. A **trinomial** contains three terms.

## Factoring

**Factoring** is the process of separating, or resolving, a quantity into factors.

No general rule can be given for factoring. In most cases the operation is performed by inspection and trial. The methods are best explained by examples.

---

If every term of a polynomial contains the same monomial factor, then that monomial is one factor of the polynomial, and the other factor is equal to the quotient of the polynomial divided by the monomial factor.

---

EXAMPLE 1: Factor the binomial $8a^2x^2 + 4a^3x$.

SOLUTION: $8a^2x^2 + 4a^3x = 4a^2x(2x + a)$

We see by inspection that $4a^2x$ is a factor common to both terms. Dividing by $4a^2x$ we arrive at the other factor.

## Exercise Set 7.2

Factor the following binomials completely.

1. $7a^2bc^3 - 28abc$
2. $9a^2x^2 - 16ax^2$
3. $5x - 25x^2$
4. $49x^4 - 16y^2$
5. $16a^4b^3 + 4a^3b$
6. $33x^3y^5 - 11x^2z$
7. $21x^2y^2 - 7x^3y^3$
8. $x^3 - 8x^2$
9. $a^2 + ab^3$
10. $6a^5 - 3ab^2$

## 7.3 Factoring Trinomials When the Coefficient of the Squared Term Equals 1

---

If a trinomial has the form

$x^2 + bx + c$

and is factorable into two binomial factors, the first term of each factor will be $x$; the second term of the binomials will be two numbers whose product is $c$ and whose sum is equal to $b$, which is the coefficient of the middle term of the trinomial.

EXAMPLE 1: Factor $x^2 + 10x + 24$.

SOLUTION: $x^2 + 10x + 24 = (x + 6)(x + 4)$

We are required to find two numbers whose product is 24 and whose sum is 10. The following pairs of factors will produce 24: 1 and 24, 2 and 12, 3 and 8, 4 and 6. From among these we select the pair whose sum is 10.

EXAMPLE 2: Factor $x^2 - 16x + 28$.

SOLUTION: $x^2 - 16x + 28 = (x - 14)(x - 2)$

We are required to find two numbers whose product is 28 and whose algebraic sum is $-16$. Since their product is positive they must both have the same sign, and since their sum is negative they must both be negative. The negative factors that will produce 28 are $-1$ and $-28$, $-2$ and $-14$, $-4$ and $-7$. We select the pair whose algebraic sum is $-16$.

EXAMPLE 3: Factor $x^2 + 5x - 24$.

SOLUTION: $x^2 + 5x - 24 = (x + 8)(x - 3)$

We are required to find two numbers whose product is $-24$ and whose algebraic sum is 5. Since their product is negative the numbers must have unlike signs, and since their sum is $+5x$, the larger number must be positive. The pairs of numbers that will produce 24, without considering signs, are 1 and 24, 2 and 12, 3 and 8, 4 and 6. From these we select the pair whose difference is 5. This is 3 and 8. We give the plus sign to the 8 and the minus sign to the 3.

EXAMPLE 4: Factor $x^2 - 7x - 18$.

SOLUTION: $x^2 - 7x - 18 = (x - 9)(x + 2)$

We are required to find two numbers whose product is $-18$ and whose algebraic sum is $-7$. Since their product is negative the signs of the two numbers are unlike, and since their sum is negative, the larger number must be negative. The pairs of numbers that will produce 18, without considering signs, are 1 and 18, 2 and 9, 3 and 6. We select the pair whose difference is 7, giving the minus sign to the 9 and the plus sign to the 2.

EXAMPLE 5: Factor $x^2 - 7xy + 12y^2$.

SOLUTION: $x^2 - 7xy + 12y^2 = (x - 4y)(x - 3y)$

We are required to find two terms whose product is $12y^2$ and whose algebraic sum is $-7y$. Since their product is positive and their sum negative they must both be negative terms. From the pairs of negative terms that will produce $+12y^2$ we select $-4y$ and $-3y$ as fulfilling the requirements.

## Exercise Set 7.3

Factor the following trinomials.

1. $a^4 + a^2 + 1$ (Hint: $a^4 = (a^2)^2$. Add and subtract $a^2$.)
2. $x^2 + 10x + 21$
3. $x^2 - 18x + 45$
4. $x^2 + 5x - 36$
5. $x^2 - 13x - 48$
6. $x^2 - 14xy + 33y^2$

7. $x^4 - 2x^2b^2 + b^4$
8. $x^2 - 12x + 36$
9. $z^4 + 16z^2 + 64$
10. $b^2 + 2bc + c^2$

## 7.4 Factoring Trinomials When the Coefficient of the Squared Term Does Not Equal 1

> If a trinomial contains three terms two of which are squares and if the third term is equal to plus or minus twice the product of the square roots of the other two, the expression may be recognized as the square of a binomial.
>
> $a^2x^2 + 2acx + c^2 = (ax + c)^2$

EXAMPLE 1: Factor $9x^2 - 24xy + 16y^2$.

SOLUTION: $9x^2 - 24xy + 16y^2$

$$= (3x)^2 - 2 \times 3x \times 4y + (4y)^2$$
$$= (3x - 4y)(3x - 4y)$$
$$= (3x - 4y)^2$$

> If an expression represents the difference between two squares, it can be factored as the product of the sum of the roots by the difference between them.
>
> $x^2 - y^2 = (x - y)(x + y)$

EXAMPLE 2: Factor $4x^2 - 9y^2$.

SOLUTION: $4x^2 - 9y^2$

$$= (2x)^2 - (3y)^2$$
$$= (2x - 3y)(2x + 3y)$$

If the factors of an expression contain like terms, these should be collected so as to present the result in the simplest form.

EXAMPLE 3: Factor $(5a + 3b)^2 - (3a - 2b)^2$.

SOLUTION: $(5a + 3b)^2 - (3a - 2b)^2$

$$= [(5a + 3b) + (3a - 2b)]$$
$$[(5a + 3b) - (3a - 2b)]$$
$$= (5a + 3b + 3a - 2b)$$
$$(5a + 3b - 3a + 2b)$$
$$= (8a + b)(2a + 5b)$$

> A trinomial in the form
> $a^4 + a^2b^2 + b^4$
> can be written in the form of the difference between two squares.

EXAMPLE 4: Resolve $9x^4 + 26x^2y^2 + 25y^4$ into factors.

SOLUTION:
$$\begin{array}{l} 9x^4 + 26x^2y^2 + 25y^4 \\ \underline{+\ 4x^2y^2 \qquad\qquad -\ 4x^2y^2} \\ (9x^4 + 30x^2y^2 + 25y^4) - 4x^2y^2 \end{array}$$

$$= (3x^2 + 5y^2)^2 - 4x^2y^2$$
$$= (3x^2 + 5y^2 + 2xy)(3x^2 + 5y^2 - 2xy)$$
$$= (3x^2 + 2xy + 5y^2)(3x^2 - 2xy + 5y^2)$$

We note that the given expression is nearly a perfect square. We therefore add $4x^2y^2$ to it to make it a square and also subtract from it the same quantity. We then write it in the form of a difference between two squares. We resolve this into factors and rewrite the result so as to make the terms follow in the order of the powers of $x$.

When a trinomial factorable into two binomials has the form $ax^2 \pm bx \pm c$, it is resolved into factors by a process of trial and error which is continued until values are found that satisfy the requirements.

EXAMPLE 5: Factor $4x^2 + 26x + 22$

SOLUTION: $4x + 11$

$$11x + 8x = 19x$$

$x + 2$          (reject)

$x + 11$

$$44x + 2x = 46x$$

$4x + 2$          (reject)

$2x + 11$

$$22x + 4x = 26x$$

$2x + 2$          (correct)

$\therefore 4x^2 + 26x + 22$

$$= (2x + 11)(2x + 2)$$

We use cross multiplication to find the required binomials. We consider the pairs of terms that will produce the first and last terms of the trinomials. We write down the various forms of examples that can be worked out with these, and we reject one trial result after another until we find the arrangement that will give us the correct value for the middle term of the given trinomial.

EXAMPLE 6: Reduce $\dfrac{12x^2 + 15x - 63}{4x^2 - 31x + 42}$.

SOLUTION:

$$\frac{12x^2 + 15x - 63}{4x^2 - 31x + 42} = \frac{(3x + 9)(4x - 7)}{(x - 6)(4x - 7)}$$

$$= \frac{3x + 9}{x - 6}$$

The numerator and denominator are factored, and the common factor is then divided out.

A fraction may be reduced to an integral or mixed expression if the degree (power) of its numerator equals or exceeds that of its denominator.

To reduce a fraction to an integral or mixed expression, divide the numerator by the denominator.

EXAMPLE 7: Reduce $\dfrac{x^2 - y^2}{x - y}$ to an integral expression.

SOLUTION:

$$\frac{x^2 - y^2}{x - y} = \frac{(x - y)(x + y)}{x - y} = x + y$$

EXAMPLE 8: Reduce $\dfrac{x^2 + y^2}{x + y}$ to a mixed expression.

SOLUTION:

$$\frac{x^2 + y^2}{x + y} = \frac{(x^2 - y^2) + 2y^2}{x + y}$$

$$= \frac{(x + y)(x - y) + 2y^2}{x + y}$$

$$= x - y + \frac{2y^2}{x + y}$$

While $x^2 + y^2$ is not evenly divisible by $x + y$, we recognize that it would be so divisible if it were $x^2 - y^2$. Therefore, we subtract $2y^2$ to convert it to $x^2 - y^2$ and also add to it the same amount. We divide $x^2 - y^2$ by $x + y$ and write the remainder as a fraction that has $x + y$ for its denominator.

EXAMPLE 9: Multiply $\dfrac{2(x + y)}{x - y}$ by $\dfrac{x^2 - y^2}{x^2 + 2xy + y^2}$.

SOLUTION:

$$\left[\frac{2(x + y)}{x - y}\right]\left[\frac{x^2 - y^2}{x^2 + 2xy + y^2}\right]$$

$$=\frac{2(x + y)(x + y)(x - y)}{(x - y)(x + y)^2} = 2$$

---

Division by a fraction is equivalent to multiplication by the reciprocal of the fraction.

---

EXAMPLE 10: Divide $\dfrac{3a^2}{a^2 - b^2}$ by $\dfrac{a}{a + b}$.

SOLUTION:

$$\frac{3a^2}{a^2 - b^2} \div \frac{a}{a + b} = \frac{3a^2}{a^2 - b^2} \cdot \frac{a + b}{a}$$

$$=\frac{3a^2(a + b)}{a(a + b)(a - b)}$$

$$=\frac{3a^2}{a(a - b)}$$

$$=\frac{3a}{a - b}$$

## Exercise Set 7.4

Factor the following trinomials.

1. $15a^2cd + 20a^2d - 15acd^2$
2. $4x^2 + 12xy + 9y^2$
3. $9a^2b^2 - 24a^2bc + 16a^2c^2$
4. $(2x + y + z)^2 - (x - 2y + z)^2$
5. $6x^2 + 21x + 9$
6. $15x^2 - 6x - 21$
7. $12x^2 + 27x - 39$
8. $9a^2 - 24ab + 16b^2$
9. $9x^4 - 12x^2 + 4$
10. $4x^2 - 20xy + 25y^2$
11. Reduce $\dfrac{x^2 + 2ax + a^2}{3(x^2 - a^2)}$ to its lowest terms.
12. Reduce $\dfrac{x^2 + a^2 + 3 - 2ax}{x - a}$ to a mixed quantity.
13. Reduce $\dfrac{x}{1 - x}$, $\dfrac{x^2}{(1 - x)^2}$, and $\dfrac{x^3}{(1 - x)^3}$ to fractions with the LCD.
14. Multiply $\dfrac{2}{x - y}$ by $\dfrac{x^2 - y^2}{a}$.
15. Divide $\dfrac{(x + y)^2}{x - y}$ by $\dfrac{x + y}{(x - y)^2}$.

## Chapter 7 Glossary

**Binomial**  A polynomial with two terms.
**Cube Root**  One of three equal factors, which, when multiplied together, produce a number.
**Factor**  An exact divisor of a number.
**Factoring**  The process of separating, or resolving, a quantity into factors.
**Fractional Exponent**  When the exponent is a fraction with a numerator of 1,

the denominator indicates the root that is to be taken of the number. When the numerator is greater than 1, the numerator indicates the power to which the number is to be raised, while the denominator indicates the root that is to be taken.

**Index**   A positive integer that indicates the root that is to be taken.

**Monomial**   An algebraic expression with one term.

**Perfect Square**   A whole number with integer square roots.

**Polynomial**   An algebraic expression with more than one term.

**Root**   An equal factor of a number.

**Square Root**   One of two equal factors, which, when multiplied together, produce a number.

**Trinomial**   A polynomial with three terms.

## Chapter 7 Test

For each problem, five answers are given. Only one answer is correct. After you solve each problem, check the answer that agrees with your solution.

1.  Which number is a root of 289?

    **A)** 7          **D)** 21
    **B)** 11         **E)** 13
    **C)** 17

2.  Which number is a root of 27?

    **A)** 12         **D)** 9
    **B)** 11         **E)** 3
    **C)** 7

3.  Which number is a root of 343?

    **A)** 9          **D)** 21
    **B)** 7          **E)** 13
    **C)** 33

4.  Which number is a root of 125?

    **A)** 17         **D)** 11
    **B)** 30         **E)** 12
    **C)** 5

5.  Find the square root of 169.

    **A)** 19         **D)** 49
    **B)** 43         **E)** 13
    **C)** 23

6.  Evaluate $\sqrt{441}$.

    **A)** 19         **D)** 11
    **B)** 21         **E)** 20
    **C)** 27

7.  Evaluate $\sqrt{1,225}$.

    **A)** 45         **D)** 15
    **B)** 26         **E)** 60
    **C)** 35

8.  Evaluate $\sqrt{25a^2}$.

    **A)** 5          **D)** $5a$
    **B)** $2a$        **E)** $5a^2$
    **C)** 25

9.  Factor $25x + 65y$ completely.

    **A)** $25(x + 15y)$
    **B)** $5(5x + 13y)$
    **C)** $25(x + y)$
    **D)** $5(5x + y)$
    **E)** $5(x + y)$

**10.** Factor $16t^2 + 14t$ completely.

   **A)** $2(8t^2 + 7t)$
   **B)** $2(t^2 + 7t)$
   **C)** $7t(2t^2 + 2)$
   **D)** $4t(4t + 3)$
   **E)** $2t(8t + 7)$

**11.** Factor $x^2 + 2x + 1$ completely.

   **A)** $(x - 1)(x + 1)$
   **B)** $(x + 1)(x + 1)$
   **C)** $(x + 2)(x - 1)$
   **D)** $(x + 1)(x + 2)$
   **E)** $(2x - 1)(x + 1)$

**12.** Factor $15x^3 + 5x^2$ completely.

   **A)** $5x(3x + 5)$
   **B)** $x^2(15x + 5)$
   **C)** $5x^2(3X + 1)$
   **D)** $15x^2(x^2 + 1)$
   **E)** $5x(3x^2 + x)$

**13.** Factor $7x - 14y$ completely.

   **A)** $2(x - 7y)$
   **B)** $7x(1 - 2y)$
   **C)** $7(x - 2y)$
   **D)** $7xy(1 - 2y)$
   **E)** $7y(x - 2y)$

**14.** Factor $x^2 + 2x - 8$ completely.

   **A)** $(x - 4)(x + 2)$
   **B)** $(x + 4)(x + 4)$
   **C)** $(4x - 2)(x + 4)$
   **D)** $(x + 4)(x + 4)$
   **E)** $(x - 4)(x + 4)$

**15.** Factor $x^2 - 7x + 12$ completely.

   **A)** $(x + 2)(x + 6)$
   **B)** $(x + 7)(x - 5)$
   **C)** $(x - 2)(x - 6)$
   **D)** $(x - 3)(x - 4)$
   **E)** $(x + 12)(x - 1)$

**16.** Factor $x^2 + 5x - 14$ completely.

   **A)** $(x - 2)(x + 7)$
   **B)** $(x - 5)(x + 7)$
   **C)** $(x - 14)(x + 1)$
   **D)** $(x - 7)(x + 2)$
   **E)** $(x - 4)(x + 10)$

**17.** Factor $x^2 - 11x + 28$ completely.

   **A)** $(x + 7)(x + 4)$
   **B)** $(x + 28)(x - 11)$
   **C)** $(x - 7)(x - 4)$
   **D)** $(2x - 4)(x - 7)$
   **E)** $(x - 6)(x + 4)$

**18.** Factor $3x^2 + 12x - 63$ completely.

   **A)** $(x + 7)(2x - 8)$
   **B)** $(2x - 7)(-4x + 3)$
   **C)** $3(x + 7)(x - 3)$
   **D)** $-3(x - 4)(x + 17)$
   **E)** $3(x + 7)(x - 5)$

**19.** Factor $-2x^2 - 6x - 4$ completely.

   **A)** $(2x - 1)(x + 5)$
   **B)** $(4x - 2)(-2x - 7)$
   **C)** $(x + 4)(x - 1)$
   **D)** $-2(x + 1)(x + 2)$
   **E)** $(x + 6)(-2x - 2)$

**20.** Factor $6x^2 - 12x + 6$ completely.

   **A)** $6(x - 1)(x + 2)$
   **B)** $-6(-x + 4)(x + 2)$
   **C)** $6(x - 1)(x - 1)$
   **D)** $3(2x - 1)(x - 2)$
   **E)** $2(3x + 2)(x + 1)$

# Linear Equations

## 8.1 Ordered Pairs

Number lines help us to graph positive and negative points on a line. If we would like to graph a line on a plane, we need to use the **Cartesian coordinate system.**

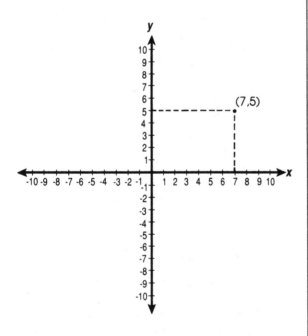

We use ordered pairs to designate the place for each point on a graph. The first number in an ordered pair is the **x-value.** The second number in an ordered pair is the **y-value.**

To plot the ordered pair (7,5), first look at the x-value, which is 7. Find positive 7 on the x-axis. Next, look at the y-value, which is 5. From +7 on the x-axis, move up on the Cartesian coordinate plane until we are next to the tick mark for +5 on the y-axis. Place a point here. The point should be where a line leading from the +7 on the x-axis meets a line leading from +5 on the y-axis.

EXAMPLE 1: Plot (2,1).

SOLUTION: Since the ordered pair is (2,1), first find the x-value, or 2, on the x-axis. Move up until we are next to the y-value, or 1, on the y-axis. Place a point here.

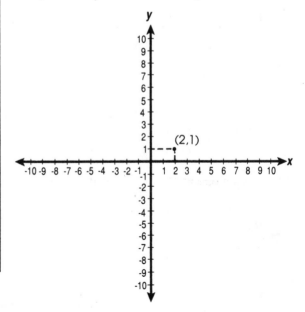

EXAMPLE 2: Plot $(-2,3)$.

SOLUTION: The ordered pair is $(-2,3)$, so find the $x$-value, or $-2$, on the $x$-axis. Next, move up to the $y$-value, or 3. Place a point here.

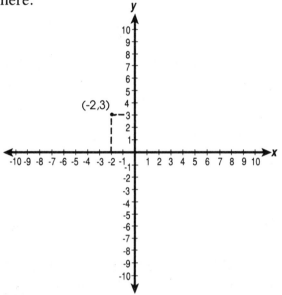

EXAMPLE 3: Plot $(-3,-2)$.

SOLUTION: Since the ordered pair is $(-3,-2)$, find the $x$-value, or $-3$, on the $x$-axis. Next, move down to the $y$-value, or $-2$. Place a point here.

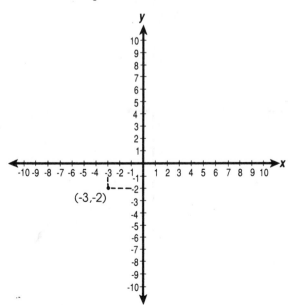

EXAMPLE 4: Plot $(5,-1)$.

SOLUTION: The ordered pair is $(5,-1)$, so find the $x$-value, or 5, on the $x$-axis. Next, move down to the $y$-value, or $-1$.

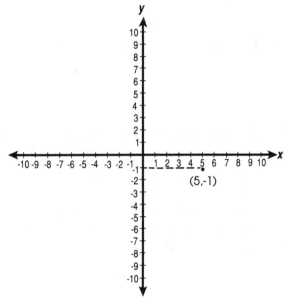

## Exercise Set 8.1

Find the $x$- and $y$-values for the points on the graph.

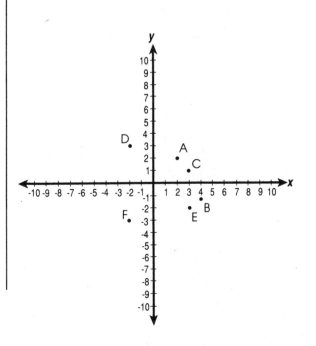

1. Point A
2. Point B
3. Point C

4. Point D
5. Point E
6. Point F

Plot these points.

7. (4,5)
8. (−2,4)

9. (3,−4)
10. (7,−7)

## 8.2 Graphing Linear Equations

To find the graph for a straight line, we can plot two points that are on the line. To find two points to plot, first we simplify the equation. Then we substitute in a value for $x$ and solve for $y$.

EXAMPLE 1: Graph $2x + y = 4$.

SOLUTION: Solve the equation for $y$, and substitute in 2 for $x$.

$2x + y = 4$

$y = 4 - 2x$    Simplify.

$y = 4 - 2(2)$    Substitute 2 for $x$.

$y = 4 - 4$

$y = 0$

The ordered pair is (2,0), and we plot it on the coordinate plane. Next, try 4 as an $x$-value.

$y = 4 - 2(4)$

$y = 4 - 8$

$y = -4$

The second ordered pair is (4,−4), and we plot this point on the coordinate plane. We draw a straight line containing both of the plotted points. This is the graph for the equation $2x + y = 4$.

The graph for any linear equation is a straight line. Since two points define a line, we only need to find two points to graph a line. However, we usually find and plot three points as a check.

For the third point, or our checkpoint, we let $x = 0$. Then we solve for $y$.

$y = 4 - 2(0)$

$y = 4 - 0$

$y = 4$

The third ordered pair is (0,4). Since when we plot this point, we find it lies on the line we have drawn, we know we have plotted our line correctly.

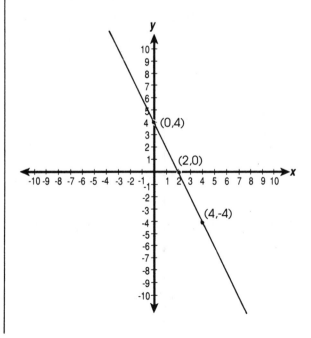

## Exercise Set 8.2

Graph each line.

1. $2x - 3 = y$
2. $x + y = 2$
3. $2y - x = 5$
4. $3y + 2x = 6$
5. $4x - y = 1$

## 8.3 Slope

The **slope** of a line tells us exactly how a line slants. Slope can be described as the ratio of the vertical change to the horizontal change between two points plotted on a line.

---

Slope = rise/run

= change in $y$/ change in $x$ = $\dfrac{y_2 - y_1}{x_2 - x_1}$

---

The slope ratio is written as $m$ in standard notation.

EXAMPLE 1: Find the slope for the equation $2x + 4 = y$.

SOLUTION: First, find two points on the line. Let $x$ equal 2, and substitute into the equation.

$$2x + 4 = y$$
$$2(2) + 4 = y$$
$$4 + 4 = y$$
$$8 = y$$

Our first ordered pair is (2,8).
Let $x$ equal 0, and substitute into the equation.

$$2(0) + 4 = y$$
$$4 = y$$

The second ordered pair is (0,4).
Use the formula for slope to find the slope, or $m$.

$$m = \frac{y_2 - y_1}{x_2 - x_1}$$
$$= \frac{4 - 8}{0 - 2}$$
$$= \frac{-4}{-2} = 2$$

The slope of the equation is 2. It does not make any difference which point is $(x_1, y_1)$, and which is $(x_2, y_2)$, as the slope will be the same.

---

A **horizontal line** has zero slope.

---

EXAMPLE 2: Find the slope for $y = 4$.

SOLUTION: For this equation, $y$ has only one value, 4. But $x$ can be equal to any value. Therefore, we have a horizontal line as a graph.

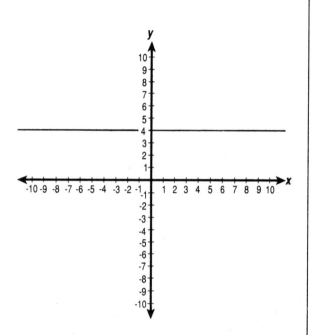

If we pick any two points on the line, we can show that the slope is zero. Two possible points are (2,4) and (7,4).

$$m = \frac{y_2 - y_1}{x_2 - x_1}$$

$$= \frac{4 - 4}{2 - 7}$$

$$= \frac{0}{-5}$$

$$= 0$$

A **vertical line** has no slope.

EXAMPLE 3: Find the slope for $x = 5$.

SOLUTION: This equation has only one $x$-value, 5. However, $y$ can have any value. Therefore, we have a vertical line as a graph.

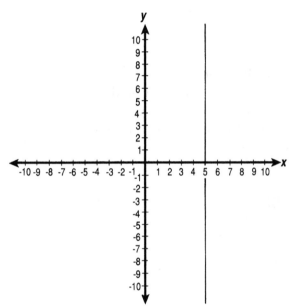

If we pick any two points on the line, we can show that there is no slope. Two possible points are (5,2) and (5,5).

$$m = \frac{y_2 - y_1}{x_2 - x_1}$$

$$= \frac{2 - 5}{5 - 5}$$

$$= \frac{-3}{0}$$

Since division by zero is undefined, the slope is undefined. Therefore, there is no slope for a vertical line.

## PRACTICALLY SPEAKING 8.3

Martha read an article in a travel magazine that stated that the distance between New York and Washington, D.C., is 4,500 miles. This distance seems much too large, as Martha knows the diameter of the earth is about 8,000 miles.

N.Y.C

D.C.

1. Is it reasonable to believe that 4,500 miles is the correct distance between New York and Washington, D.C.?

Martha looked up the distance on her Rand Mc-Nally map, and found that the distance was much less than 4,500 miles. Martha is puzzled, as she cannot understand why the magazine would print the wrong number.

2. What could have happened?

See Appendix F for the answers.

## Exercise Set 8.3

Find the slope of each linear equation.

1. $2x - 7 = y$
2. $3y + x = 4$
3. $6 - 2y = 4x$
4. $x - y = 5$
5. $y - 4 = 5x$
6. $2x + 2y = 4$
7. $y = 7$
8. $2x = 3$
9. $3x = -5$
10. $3y = 12$

## 8.4 Systems of Linear Equations

**Simultaneous linear equations,** or **linear systems of equations,** are equations that contain the same variables. Thus, $a + 2b - 11$ and $2a + b = 10$ are simultaneous equations since they both involve the same variables, $a$ and $b$.

Simultaneous equations involving two variables are solved by using one of three methods. That is, we use one of these methods to find the values for the variables that satisfy both the equations in the system.

Methods for Solving Linear Systems of Equations

Method 1: Elimination

Method 2: Substitution

Method 3: Comparison

## 8.5 Solving Linear Systems of Equations by Elimination

We can solve systems of equations by using addition or subtraction. This is known as the **elimination** method.

**Step 1.** Multiply one or both of the equations by such a number or numbers as will give one of the variables the same coefficient in both equations.

**Step 2.** Add or subtract the equal coefficients according to the nature of their signs.

EXAMPLE 1: $5x + 2y = 32, 2x - y = 2$. Find $x$ and $y$.

SOLUTION:

$$5x + 2y = 32 \quad \text{Multiply } 2x - y$$
$$\underline{4x - 2y = 4} \quad \text{by 2.}$$
$$9x \quad = 36$$
$$x \quad = 4$$
$$20 + 2y = 32 \quad \text{Substitute 4 for } x$$
$$\quad \text{in first equation.}$$
$$2y = 32 - 20 \quad \text{Transpose.}$$
$$y = 6$$

EXAMPLE 2: Solve this system using the elimination method.

$$4x - 2 = 6y$$
$$2x - 1 = 3y$$

SOLUTION:

$$4x - 2 = 6y \quad \text{Multiply } 2x - 1 = 3y$$
$$\underline{4x - 2 = 6y} \quad \text{by 2.}$$
$$0 = 0 \quad \text{Subtract.}$$

Since the equations are identical when we make the coefficients of one of the variables equal, our result when we try to eliminate one of the variables is $0 = 0$. This means that there are an infinite number of values of $x$ and $y$ that are solutions for both the equations. We call a system with an infinite number of solutions a **dependent system.**

EXAMPLE 3: Solve this system using the elimination method.

$$y = 2x + 4$$
$$y = 2x - 3$$

SOLUTION:

$$y = 2x + 4$$
$$\underline{y = 2x - 3}$$
$$0 = \quad 7$$

When we subtract to solve for one of the variables, we find that the result is $0 = 7$. Since this is false, there is no solution possible for this system of equations. We call a system with no solutions an **inconsistent system.**

## Exercise Set 8.5

Use the elimination method to solve each system of linear equations.

1. $x + y = 6$
   $2x - y = 3$
2. $2x + 3y = 6$
   $5x - 4y = -8$
3. $y = -x + 4$
   $2y = -2x + 8$
4. $x - 8 = 3y$
   $x + 14 = 3y$
5. $2y = 4x - 3$
   $6y = 12x + 6$
6. $2x - y = 4$
   $4x + y = 14$

7. The hands of a clock are together at 12 o'clock. When do they next meet ($x$ = minute spaces passed over by minute hand; $y$ = number passed over by hour hand)?

8. There are two numbers: the first added to half the second gives 35; the second added to half the first equals 40. What are the two numbers?

9. Janet and Tim invest $918 in a partnership venture and earn $153. Janet's share of the profit is $45 more than Tim's. How much did each contribute if their profits are proportional to their investments?

## 8.6 Solving Systems of Linear Equations by Substitution or Comparison

We can also solve systems of linear equations by using the **substitution method.**

**Step 1.** From one of the equations find the value of one of the variables in terms of the other.

**Step 2.** Substitute the value thus found for the variable in the other of the given equations.

EXAMPLE 1: $2x + 4y = 50$, $3x + 5y = 66$. Find $x$ and $y$.

SOLUTION:

$$2x + 4y = 50$$
$$2x = 50 - 4y \quad \text{Transpose.}$$
$$x = 25 - 2y$$

$$3(25 - 2y) + 5y = 66 \quad \text{Substitute for } x \text{ in other equation.}$$

$$75 - 6y + 5y = 66$$
$$-y = 66 - 75 = -9, y = 9.$$

$$2x + 36 = 50 \quad \text{Substitute 9 for } y \text{ in first equation.}$$

$$2x = 50 - 36 = 14$$
$$x = 7$$

Our third method of solving linear equations is called the **comparison method.** When solving a system of two linear equations that involve two of the same variables, we first solve both equations for one of the variables. Then we set the equations equal to one another.

**Step 1.** From each equation find the value of one of the variables in terms of the other.

**Step 2.** Form an equation from these equal values.

EXAMPLE 2: $3x + 2y = 27$, $2x - 3y = 5$. Find $x$ and $y$.

SOLUTION:

$$3x + 2y = 27$$
$$3x = 27 - 2y$$
$$x = \frac{27 - 2y}{3}$$

$$2x - 3y = 5$$
$$2x = 5 + 3y$$
$$x = \frac{5 + 3y}{2}$$

$$\frac{27 - 2y}{3} = \frac{5 + 3y}{2} \quad \text{Both are equal to } x.$$

$$27 - 2y = \frac{3(5 + 3y)}{2}$$  Multiply both sides by 3.

$$2(27 - 2y) = 3(5 + 3y)$$  Multiply both sides by 2.

$$54 - 4y = 15 + 9y$$  Carry out multiplication.

$$-4y - 9y = 15 - 54$$
$$= -39$$
$$y = 3$$
$$3x + 6 = 27$$
$$3x = 21$$
$$x = 7$$

Of the foregoing methods, select the one which appears most likely to make the solution simple and direct.

## Exercise Set 8.6

Solve each system of linear equations using either the substitution or the comparison method.

1. $x = 3$
   $x + 2y = 7$
2. $x + 2y = 14$
   $x = -6$
3. $x + 3y = 6$
   $4y = 12$
4. $x + y = 4$
   $x = 1$
5. $x + 4y = 10$
   $y = -2$

6. Alan has $22,000 invested, and he earns $1,220 on his investment. Part of the money is in a money market fund earning 6% interest, and part is in a savings account earning 5%. How much is invested each way?

7. Jack is twice as old as Joe. Twenty years ago Jack was four times as old as Joe. What are their ages now?

8. The inventory of one department in Jason's Department Store increased by ⅓ of that of a second department amounts to $1,700. The inventory of the second increased by ¼ of that of the first amounts to $1,800. What are the inventories of each department?

9. Find two numbers such that ½ of the first plus ⅓ of the second shall equal 45, and ½ of the second plus ⅕ of the first shall equal 40.

10. Two girls receive $153 for baby-sitting. Ann is paid for 14 days and Mary for 15. Ann's pay for 6 days' work is $3 more than Mary gets for 4. How much do they each earn per day?

11. In 80 pounds of an alloy of copper and tin, there is a ratio of 7 pounds of copper to 3 pounds of tin. How much copper must be added so that there is a ratio of 11 pounds of copper to 4 of tin?

12. Brian owes $1,200 and Jamie $2,500, but neither has enough money to pay his debts. Brian says to Jamie, "Lend me ⅛ of your savings account, and I'll pay off all my debts." Jamie says to Brian, "Lend me ⅑ of yours and I'll pay all of mine." How much money does each one have?

## Chapter 8 Glossary

**Cartesian Coordinate System**   Two perpendicular number lines used to place points in a plane.

**Comparison Method**   A method of solving linear systems of equations by solving two equations for the same variable, and then setting the results equal to one another.

**Dependent System**   A system with an infinite number of solutions.

**Elimination Method**   A method of solving linear systems of equations using addition or subtraction.

**Horizontal Line**   A line with zero slope.

**Inconsistent System**   A system with no solutions.

**Linear Systems of Equations**   Two or more linear equations that contain the same variables.

**Slope**   The ratio of the vertical change to the horizontal change between two points plotted on a line.

**Substitution Method**   A method of solving linear systems of equations by solving for one variable in terms of another, and then substituting in the value of the first variable to solve for the second variable.

**Vertical Line**   A line with no slope.

***x*-value**   The first coordinate in an ordered pair.

***y*-value**   The second coordinate in an ordered pair.

## Chapter 8 Test

For each problem, five answers are given. Only one answer is correct. After you solve each problem, check the answer that agrees with your solution.

Use the graph to answer questions 1 through 3.

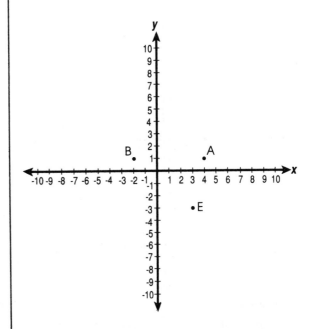

1. What are the coordinates for point *A*?

   **A)** $(-1,-3)$
   **B)** $(1,4)$
   **C)** $(4,1)$
   **D)** $(0,5)$
   **E)** $(-4,-1)$

2. What are the coordinates for point *E*?

   **A)** $(-1,-3)$
   **B)** $(4,1)$
   **C)** $(2,4)$
   **D)** $(3,-3)$
   **E)** $(-2,1)$

3. What are the coordinates for point *B*?

   **A)** $(3,1)$
   **B)** $(-2,1)$
   **C)** $(-1,-2)$
   **D)** $(1,3)$
   **E)** $(4,1)$

**4.** What is the equation for this graph?

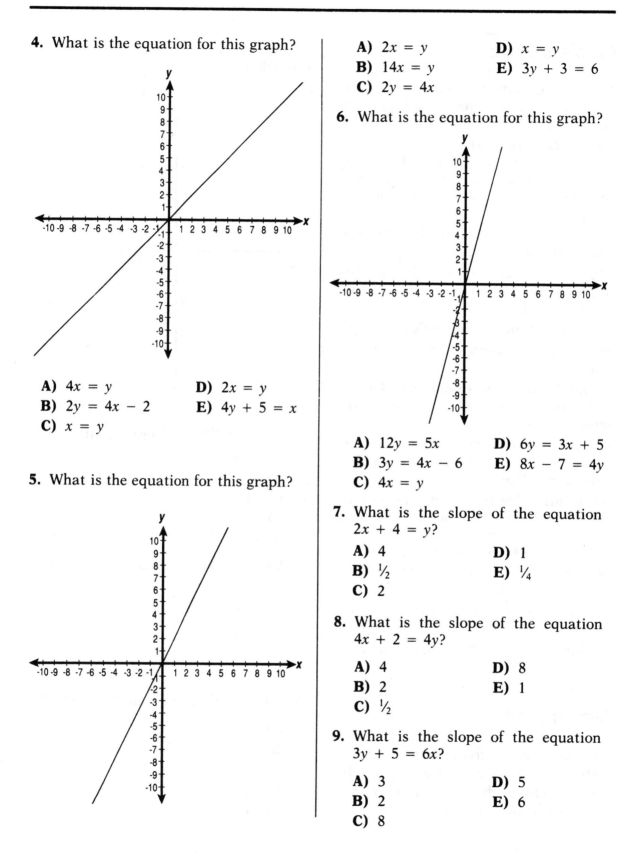

**A)** $4x = y$        **D)** $2x = y$

**B)** $2y = 4x - 2$   **E)** $4y + 5 = x$

**C)** $x = y$

**5.** What is the equation for this graph?

**A)** $2x = y$        **D)** $x = y$

**B)** $14x = y$       **E)** $3y + 3 = 6$

**C)** $2y = 4x$

**6.** What is the equation for this graph?

**A)** $12y = 5x$      **D)** $6y = 3x + 5$

**B)** $3y = 4x - 6$   **E)** $8x - 7 = 4y$

**C)** $4x = y$

**7.** What is the slope of the equation $2x + 4 = y$?

**A)** 4              **D)** 1

**B)** ½             **E)** ¼

**C)** 2

**8.** What is the slope of the equation $4x + 2 = 4y$?

**A)** 4              **D)** 8

**B)** 2              **E)** 1

**C)** ½

**9.** What is the slope of the equation $3y + 5 = 6x$?

**A)** 3              **D)** 5

**B)** 2              **E)** 6

**C)** 8

**10.** What is the slope of the equation $2y + 17 = 8x$?

**A)** 8          **D)** $\frac{1}{4}$

**B)** 2          **E)** 17

**C)** 4

**11.** Solve the system of linear equations for $x$.

$$y = 5$$
$$2y + 3x = 4$$

**A)** 2          **D)** $-2$

**B)** $-6$          **E)** $-5$

**C)** 4

Use this system of linear equations to answer questions 12 and 13.

$$2x + 3y = 6$$
$$5x - 4y = -8$$

**12.** Solve for $x$.

**A)** 2          **D)** 4

**B)** 0          **E)** 5

**C)** 6

**13.** Solve for $y$.

**A)** 4          **D)** 2

**B)** $-8$          **E)** $-4$

**C)** 0

Use this system of linear equations to answer questions 14 and 15.

$$3x + 5y = 6$$
$$2x + y = 4$$

**14.** Solve for $x$.

**A)** 2          **D)** 4

**B)** 0          **E)** 5

**C)** $-1$

**15.** Solve for $y$.

**A)** 4          **D)** 0

**B)** $-1$          **E)** 3

**C)** 2

Use this system of linear equations to answer questions 16 and 17.

$$x = 2y$$
$$x + y = 12$$

**16.** Solve for $x$.

**A)** 4          **D)** 8

**B)** 1          **E)** 12

**C)** 2

**17.** Solve for $y$.

**A)** 2          **D)** 4

**B)** 1          **E)** 3

**C)** 6

Use this system of linear equations to answer questions 18 and 19.

$$x + y = 3$$
$$x - y = 1$$

**18.** Solve for $x$.

**A)** 2          **D)** $\frac{1}{2}$

**B)** 1          **E)** 4

**C)** 3

**19.** Solve for $y$.

**A)** 2          **D)** 4

**B)** 1          **E)** 3

**C)** $\frac{1}{2}$

**20.** What is the slope of the equation $5x + y = 22$

**A)** 22          **D)** 1

**B)** 11          **E)** $-5$

**C)** 5

# Series

## 9.1 Number Series

A **number series,** or progression, is a sequence of numbers arranged according to a pattern.

A series is **ascending** if the numbers are increasing from beginning to end. A series is **descending** if the numbers are decreasing from the beginning to the end of the series.

EXAMPLE 1: Is the series 1, 3, 5, 7, 9 . . . ascending or descending?

SOLUTION: The series is ascending, as the numbers get larger as the series progresses.

EXAMPLE 2: Is the series 31, 27, 23, 19 . . . ascending or descending?

SOLUTION: The series is descending, as the numbers get smaller as the series progresses.

## Exercise Set 9.1

State whether the series are ascending or descending series.

1. 2, 4, 6, 8, 10 . . .
2. 55, 44, 33, 22, 11 . . .
3. 2, 4, 7, 9, 12 . . .
4. 12, 9, 6, 3 . . .
5. 34, 45, 56, 67, 78 . . .
6. 1, 2, 3, 4 . . .
7. 123, 124, 125, 126 . . .
8. 77, 35, 27, 15 . . .
9. 35, 30, 25, 20 . . .
10. 14, 16, 18, 20 . . .

## 9.2 Arithmetic Series

An **arithmetic series** is one in which the successive numbers are formed by addition or subtraction. The series

$$1, 3, 5, 7 \ldots$$

is formed by adding 2 to each term to make up the next term.

To find missing terms in an ascending arithmetic series, subtract any term from the next term to get the difference, or **increment.** Then add the increment to the term in front of any missing term to obtain the missing term.

EXAMPLE 1: Supply the missing terms in 1, 4, 7, 10, 13, ____, ____.

SOLUTION: First, find the increment.

$$4 - 1 = 3$$

The difference between the missing terms is 3. So add 3 to 13 to get the first missing term.

$$13 + 3 = 16$$

Add 3 to 16 to get the next missing term.

$$16 + 3 = 19$$

The series is now: 1, 4, 7, 10, 13, 16, 19.
To find the missing terms in a descending arithmetic series, subtract any term from the preceding term to get the difference. Then subtract the difference from the term in front of any missing term to find the missing term.

EXAMPLE 2: Supply the missing terms in 14, 11, 8, 5, ____, ____.

SOLUTION: First, find the difference.

$$14 - 11 = 3$$

Now subtract the difference from the term in front of one of the missing terms.

$$5 - 3 = 2$$

Subtract 3 from 2 to get the second missing term.

$$2 - 3 = -1$$

The series is now: 14, 11, 8, 5, 2, −1.
To find a given term in an arithmetic series, first determine the difference. Then multiply the difference by the number of terms minus 1, and add the first term to the product.

EXAMPLE 3: Find the tenth term in the series:

$$3, 6, 9, 12, \underline{\quad}.$$

SOLUTION: The difference equals 3. The number of terms minus 1 = (10 − 1) = 9

$$9 \times 3 = 27$$
$$27 + 3 = 30$$

Check: Write out the series to check the answer.

$$3, 6, 9, 12, 15, 18, 21, 24, 27, 30$$

## Exercise Set 9.2

Find the missing terms in each arithmetic series.

1. 4, 6, 8, 10, ____, 14, ____
2. 21, 23, ____, 27, 29, ____
3. ____, 3, 6, 9, ____, 15, 18
4. 23, ____, 17, 14, ____, 8
5. 5, 10, ____, 20, 25, ____
6. 77, 66, ____, 44, 33, ____
7. 4, 8, 12, ____, 20, ____

8. ____, 30, 25, ____, 15, 10
9. 45, 36, ____, 18, ____
10. 14, 21, ____, ____, 42, 49

## 9.3 Geometric Series

A **geometric series** is one in which the successive terms are formed by multiplication or division.

This sequence is formed when each term, beginning with 2, is multiplied by 2 to make the next term.

$$2, 4, 8, 16, 32, 64 \ldots$$

To find missing terms in a geometric series, divide any term by the preceding term to find the multiplier, or the **ratio.** Then multiply any term by the ratio to find the next term in the series. Divide any term by the ratio to find the preceding term in the series.

EXAMPLE 1: Supply the missing terms in the series:

$$\text{\_\_\_\_}, 3, 6, 12, 24, 48, \text{\_\_\_\_}.$$

SOLUTION: First, find the ratio. Since $6 \div 3 = 2$, 2 is the ratio.

$$48 \times 2 = 96, \text{ the last term}$$

$$3 \div 2 = \frac{3}{2}, \text{ the first term}$$

The series is now: $\frac{3}{2}$, 3, 6, 12, 24, 48, 96.

EXAMPLE 2: Supply the missing terms in the series:

$$224, 112, \text{\_\_\_\_}, 28, 14, \text{\_\_\_\_}$$

SOLUTION: First, find the ratio. Since $14 \div 28 = \frac{1}{2}$, the ratio is $\frac{1}{2}$.

Now find the missing terms.

$$112 \times \frac{1}{2} = 56$$

$$14 \times \frac{1}{2} = 7$$

The series is now: 224, 112, 56, 28, 14, 7.

## Exercise Set 9.3

Find the ratio for each geometric series, and then find the missing terms.

1. 2, 4, 8, 16, 32, 64, ____, ____
2. 96, 48, 24, ____, ____
3. 2, 10, 50, 250, ____, ____
4. 243, 81, ____, 9, 3, ____
5. 3, 12, 48, ____, 768, 3,072, ____
6. 8, ____, 32, 64, 128, ____
7. 3, 9, 27, ____, ____, 729
8. 5, $5^2$, ____, ____, $5^5$
9. 24, 96, ____, ____, 6,144
10. 4, 20, ____, ____, 2,500

## 9.4 Finding Sums of Series

To find the sum of a series, we must first examine the series to see if there is a pattern. If we find a pattern to the series, that is, if we find that we can represent the series by an arithmetic or geometric formula, we can then generally find the sum of the series.

To find the sum of an arithmetic series, divide the number of terms by 2, and multiply this by the sum of the first and last term of the series.

$$S = \frac{n(a + l)}{2}$$

The **formula** $S = \frac{n(a + l)}{2}$, where $S$ is the sum, $n$ is the number of terms, $a$ is the first term, and $l$ is the last term, finds the sum of an arithmetic series.

EXAMPLE 1: What is the sum of the numbers from 1 through 10?

SOLUTION:

Substituting for $n$, $a$, and $l$ in the formula we find:

$$S = \frac{10(1 + 10)}{2} = \frac{110}{2} = 55$$

We check the answer by addition.

$$1 + 2 + 3 + 4 + 5 + 6 + 7$$
$$+ 8 + 9 + 10 = 55$$

To find the sum of a geometric series, multiply the last term by the ratio, subtract the first term from this product, and divide the remainder by the ratio minus 1.

$$S = \frac{rl - a}{r - 1}$$

EXAMPLE 2: What is the sum of the series 4, 16, 64, 256, 1,024?

SOLUTION: We use the formula for finding the sum of a geometric series, and substitute for r, l, and $a$.

$$S = \frac{(4 \times 1,024) - 4}{4 - 1}$$
$$= \frac{4,092}{3}$$
$$= 1,364$$

## Exercise Set 9.4

Find the sums of the following series.

1. 2, 4, 6, 8, 10, 12, 14
2. 21, 24, 27, 30, 33, 36, 39, 42
3. 4, 8, 12, 16, 20, . . . , 100
4. 2, 4, 8, 16, 32, 64, 128
5. 3, 9, 27, . . . , 6,561
6. 5, $5^2$, $5^3$, . . . , $5^6$
7. 12, 16, 20, 24, 28
8. 1, 5, 9, 13, 17, 21, 25
9. 6, 24, 96, . . . , 7,776
10. 4, 20, 100, . . . , 2,500

## Chapter 9 Glossary

**Arithmetic Series**  A series in which the successive numbers are formed by addition or subtraction.

**Ascending Series**  A series where the numbers are increasing from beginning to end.

**Descending Series**  A series where the numbers are decreasing from beginning to end.

**Geometric Series**   A series in which the successive numbers are formed by multiplication or division.

**Increment**   The difference between two successive terms in a series.

**Number Series**   A sequence of numbers arranged according to a pattern.

**Ratio**   The multiplier found by dividing any term by the preceding term in a geometric series.

## Chapter 9 Test

For each problem, five answers are given. Only one answer is correct. After you solve each problem, check the answer that agrees with your solution.

1. Which term listed below is missing from the series: 12, 14, _____, 18, _____ .

   **A)** 17          **D)** 19
   **B)** 16          **E)** 25
   **C)** 8

2. Which term listed below is missing from the series: 26, 23, _____, 17, _____, 11?

   **A)** 16          **D)** 13
   **B)** 8           **E)** 14
   **C)** 21

3. Which term listed below is missing from the series: 2, 4, 6, _____, 10, _____, 14?

   **A)** 12          **D)** 13
   **B)** 22          **E)** 20
   **C)** 9

4. Which term listed below is missing from the series: 25, 20, _____, 10, 5, _____?

   **A)** 36          **D)** 22
   **B)** 45          **E)** 0
   **C)** 1

5. Which term listed below is missing from the series: 6, 12, _____, 24, 36, 42?

   **A)** 40          **D)** 32
   **B)** 16          **E)** 18
   **C)** 20

6. Which term listed below is missing from the series: 4, 7, 10, _____, 16, _____, 22?

   **A)** 12          **D)** 15
   **B)** 18          **E)** 20
   **C)** 19

7. Which term listed below is missing from the series: 3, 6, _____, 24, 48, _____?

   **A)** 95          **D)** 12
   **B)** 18          **E)** 8
   **C)** 106

8. Which term listed below is missing from the series: 64, _____, 16, 8, _____, 2?

   **A)** 12          **D)** 6
   **B)** 36          **E)** 32
   **C)** 48

9. Which term listed below is missing from the series: 15, 13, _____, 9, _____, 5?

   **A)** 3           **D)** 7
   **B)** 8           **E)** 17
   **C)** 10

**10.** Which term listed below is missing from the series: 7, _____, 28, 56, _____?

A) 114          D) 21
B) 14           E) 11
C) 212

**11.** Which series is an ascending series?

A) 4, 2, −1 . . .
B) 14, 6, −2 . . .
C) 12, 10, 8, 6 . . .
D) 25, 23, 21, 19 . . .
E) 6, 12, 18, 24 . . .

**12.** Which series is a descending series?

A) 4, 8, 12, 16 . . .
B) 12, 10, 8, 6 . . .
C) 1, 3, 5, 7 . . .
D) 5, 25, 125, 625 . . .
E) 10, 20, 30, 40 . . .

**13.** Find the sum of the series: 4, 8, 12, 16, 20.

A) 75          D) 45
B) 84          E) 72
C) 60

**14.** Find the sum of the series: 3, 6, 9, 12, 15.

A) 36          D) 60
B) 42          E) 45
C) 25

**15.** Find the sum of the series: 2, 8, 32, 128.

A) 192        D) 200
B) 170        E) 188
C) 150

**16.** Find the sum of the series: 1, 3, 9, 27.

A) 35          D) 58
B) 45          E) 40
C) 52

**17.** Find the sum of the series: 12, 18, 24, 30, 36.

A) 240        D) 96
B) 120        E) 400
C) 360

**18.** Find the sum of the series: 8, 16, 32, 64, 128.

A) 256        D) 320
B) 196        E) 296
C) 248

**19.** Find the sum of the series: 5, 15, 45, 135, 405.

A) 605        D) 590
B) 575        E) 645
C) 705

**20.** Find the sum of the series: 20, 40, 80, 160, 320.

A) 580        D) 620
B) 840        E) 760
C) 460

# Geometry

## 10.1 Basic Concepts

**Geometry** is the branch of mathematics that deals with space relationships. Applications of the principles of geometry require an ability to use arithmetic and elementary algebra as shown in this book. A knowledge of geometry in addition to simple algebra and arithmetic is basic to many occupations, such as carpentry, dress design, machine-shop work, tool-making, drafting, architecture, and engineering.

A **geometric figure** is a point, line, plane, solid, or combination of these.

A **point** is the position of the intersection of two lines. It does not have length, width, or thickness.

A **line** is the intersection of two surfaces. It has length, but does not have either width or thickness. It may be straight, curved, or broken.

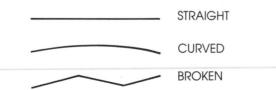

STRAIGHT

CURVED

BROKEN

In solving geometric problems we apply certain general principles that demonstrate a relationship called **theorems.** These are demonstrated by means of basic principles called **axioms** and **postulates.**

We may think of the **axioms** used in geometry as basic mathematical principles that are so elementary they cannot be demonstrated by simpler principles. They were once widely called "self-evident truths." Note that the first seven "axioms" listed below are the principles with which we have already become familiar in performing operations upon algebraic equations.

The **postulates** used in geometry are of two different, but closely related, kinds. Some are merely restatements of more general mathematical axioms in specific geometric terms. Others are axiom-like statements which apply only to geometry. For instance, the last three "axioms" below may also be thought of as geometric postulates.

### Axioms

1. Things equal to the same thing are equal to each other.

2. If equals are added to equals, the sums are equal.

3. If equals are subtracted from equals, the remainders are equal.

4. If equals are multiplied by equals, the products are equal.

5. If equals are divided by equals, the quotients are equal.

6. The whole is greater than any of its parts, and is equal to the sum of all its parts.

7. A quantity may be substituted for an equal one in an equation or in an inequality.

8. Only one straight line can be drawn through two points.

9. A straight line is the shortest distance between two points.

10. A straight line may be produced to any required length.

## 10.2 Lines, Angles, and Constructions
### Lines

A **horizontal line** is a straight line that is level with the horizon, or goes straight across from left to right.

HORIZONTAL

A **vertical line** is a straight line that goes straight up and down.

VERTICAL

Two lines are **perpendicular** to one another when the two lines intersect and form all right angles.

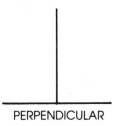

PERPENDICULAR

**Parallel lines** are two or more straight lines that are equally distant from one another at all points. Parallel lines never meet.

PARALLEL

### Angles

An **angle** is the figure formed by two lines meeting at a common point called the **vertex.** The lines that form an angle are called its **sides.** If three letters are used to describe the angle, the vertex is read between the others. Thus, the figure below is written ∠*ABC*, and is read angle *ABC*; the sides are *AB* and *BC*.

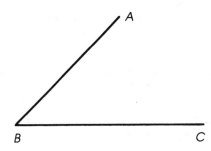

The unit of measurement for angles is the **degree.**

A **straight angle** is one of 180°. Its two sides lie on the same straight line.

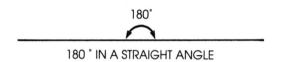

180° IN A STRAIGHT ANGLE

A **right angle** is one of 90°. It can also be described as one half of a straight angle.

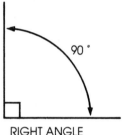

RIGHT ANGLE

An **acute angle** is any angle that is less than (<) a right angle. Thus an acute angle must measure less than 90°.

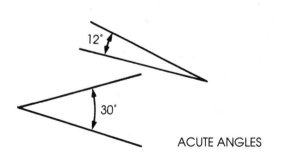

ACUTE ANGLES

An **obtuse angle** is greater than (>) a right angle, but less than (<) a straight angle. Thus, an obtuse angle must measure between 90° and 180°.

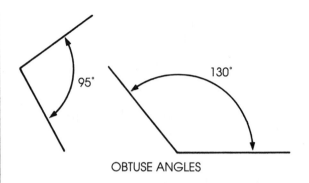

OBTUSE ANGLES

Angles are measured by determining the opening between the sides of an angle, rather than the length of the sides. To measure angles, use a protractor such as the one illustrated.

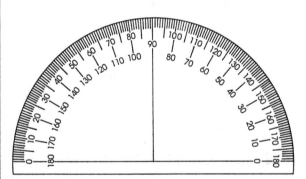

PROTRACTOR FOR MEASURING ANGLES

To measure an angle with a protractor, place the center of the protractor at the vertex of the angle, and the straight side on a line with one side of the angle. Read the degrees where the other side of the angle crosses the scale of the protractor.

To draw an angle with a protractor, draw a straight line for one side of the angle. Place the center of the protractor at the point of the line that is to be the vertex of the angle, and make the straight line of the protractor coincide with the line. Place a dot on the paper at the point on the scale of the protractor that corresponds to the size of the angle to be drawn. Connect this dot and the vertex to obtain the desired angle.

When measuring an angle, remember that we can think of it as being made up of the spokes of a wheel, where all the spokes come from a point at the center of the wheel.

## Geometric Constructions

Geometrical constructions, in the strict sense, involve only the use of a straight-edge and a compass. These are the only instruments necessary for the following constructions.

EXAMPLE 1: Bisect a straight line. (Bisect means to divide in half.)

SOLUTION: With points *A* and *B* as centers

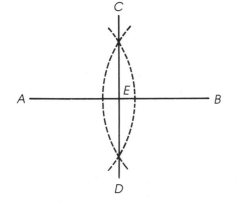

and with a radius greater than half the line *AB*, use a compass to draw arcs intersecting at points *C* and *D*. Draw *CD*, which bisects *AB* at point *E*. (It should be noted that *CD* is perpendicular to *AB*.)

EXAMPLE 2: Bisect any angle.

SOLUTION: With the vertex as center and any radius draw an arc cutting the sides of the angle at points *B* and *C*. With points *B* and *C* as centers and with a radius

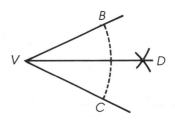

greater than half the distance from point *B* to point *C*, describe two arcs intersecting at point *D*. The line *DV* bisects ∠*CVB*.

EXAMPLE 3: At a given point on a line, construct a perpendicular to the line.

SOLUTION: From the given point *P* as center with any radius describe an arc which cuts the line *AB* at points *M* and *N*. From points *M* and *N* as centers and with a radius

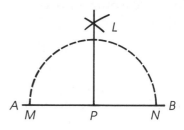

greater than *MP*, describe arcs which intersect at point *L*. Draw the line *PL*, which is the required perpendicular.

EXAMPLE 4: From a given point away from a straight line drop a perpendicular to the line.

SOLUTION: From the given point $P$ as center and with a large enough radius describe an arc which cuts line $AB$ at points $C$ and $D$. From points $C$ and $D$ as centers and with a radius greater than half $CD$, describe two

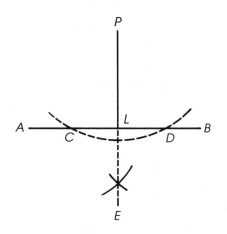

arcs that intersect at point $E$. Connect $PE$. The line $PL$ is the required perpendicular to the line $AB$.

*Note:* For some of the previous constructions and some that are to follow, more than one method is available. To avoid confusion in learning, only one method is presented here.

EXAMPLE 5: Duplicate a given angle.

SOLUTION: Let the given angle be $\angle AVB$. Then from the vertex $V$ as center and with a convenient radius, draw an arc that intersects the sides at points $C$ and $D$. Draw any straight line equal to or greater in length than $VB$ and call it $V'B'$. (Read $V$ *prime B prime.*) With point $V'$ as center and with the same radius, describe an arc point $C'E'$ that cuts the line at point $C'$. From

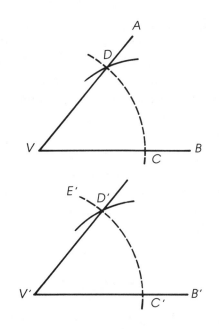

point $C'$ as center and with a radius equal to $DC$, describe an arc intersecting arc $C'E'$ at point $D'$. Draw $D'V'$. $\angle D'V'C'$, is the required angle.

EXAMPLE 6: Duplicate a given triangle.

SOLUTION: Draw any straight line from any point $D$ as center, and with a radius equal

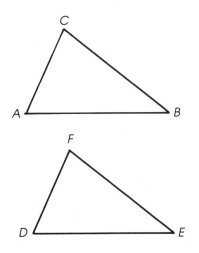

to *AB* lay off *DE* equal to *AB*. With point *E* as center and *BC* as radius, draw an arc. With point *D* as center and *AC* as radius, draw an arc which intersects the other arc at point *F*. Draw *FE* and *FD*. *DEF* is the required triangle.

EXAMPLE 7: Construct a line parallel to a given line through a given point not on the given line.

SOLUTION: If the given line is *AD*, and the given point is *P*, then draw a line *PQ* through any point *Q* on *AD*. Label the angle *PQA*. Construct a corresponding angle at *P* congruent to ∠*PQA*, using the method in EXAMPLE 5. Label the congruent angle *UPS*. Line *SP* is parallel to line *AD*.

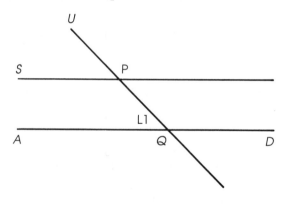

EXAMPLE 8: Divide a line into a given number of equal parts.

SOLUTION: If *AB* is the given line, and if it is to be divided into six parts, then draw line *AC* making an angle (most conveniently an acute angle) with *AB*. Starting at point *A*, mark off on *AC* with a compass six equal divisions of any convenient length. Connect the last point *I* with point *B*. Through points *D*, *E*, *F*, *G*, and *H* draw

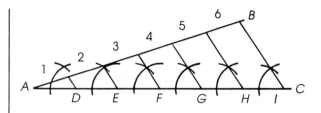

lines parallel to *IB* by making equal angles. The parallel lines divide *AB* into six equal parts.

EXAMPLE 9: Find the center of a circle or arc of a circle.

SOLUTION: Draw any two chords *AB* and *DE*. A chord is a segment whose end points lie on the circle. Draw the perpendicular bisectors of these chords. (See EXAMPLE 1.)

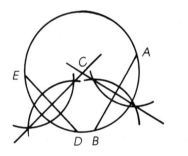

The point *C* where they intersect is the center of the circle or arc.

EXAMPLE 10: Inscribe a regular **hexagon** in a circle.

SOLUTION: *Note* that a regular hexagon is a polygon with six equal sides and six equal angles. The length of a side of a hexagon is equal to the radius of a circle circumscribing it. The radius of the circle is equal to *AG*. Starting at any point on the circle and using the length of the radius

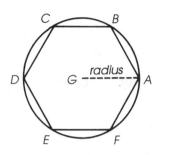

as the distance, lay off successive points *B, C, D, E, F* on the circumference of the circle. Connect the points with straight lines to obtain the required hexagon.

## Exercise Set 10.2

Use a protractor to answer the following questions.

**1.** Draw a straight line.

**2.** Draw a right angle.

**3.** Draw an obtuse angle of 30°.

**4.** Draw an obtuse angle of 120°.

Use the diagram below to answer the following questions.

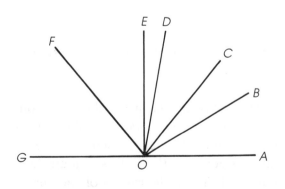

**5.** What is the measure of ∠*AOB*?
**6.** What is the measure of ∠*AOC*?

**7.** What is the measure of ∠*AOD*?
**8.** What is the measure of ∠*AOE*?
**9.** What is the measure of ∠*AOF*?
**10.** What is the measure of ∠*BOF*?
**11.** What is the measure of ∠*BOD*?

## 10.3 Line and Angle Relationships

Now that we have learned some basic geometric definitions and constructions, we will discuss some important relationships between lines and angles.

We introduce definitions, postulates, propositions, theorems, and corollaries to discuss the relationships between lines and angles.

The following are important geometric postulates.

**Postulate 1.** A geometric figure may be moved from one place to another without changing its size or shape.

**Postulate 2.** Two angles are equal if they can be made to coincide.

**Postulate 3.** A circle can be drawn with any point as center.

**Postulate 4.** Two straight lines can intersect in only one point.

**Postulate 5.** All straight angles are equal.

A **corollary** is a geometric truth that follows from one previously given and needs little or no proof.

For example, from Postulate 3 we derive the corollary:

**Corollary 1.** An arc of a circle can be drawn with any point as center.

**Adjacent angles** are angles that have a common vertex and a common side between them.

For example, ∠*CPB* is adjacent to ∠*BPA* but not to ∠*DRC*.

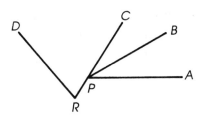

**Postulate 6.** Adjacent angles can be added. Thus:

$$\angle AOB + \angle BOC = \angle AOC$$
$$\angle DOC + \angle COB + \angle BOA = \angle DOA$$
$$\angle EOD + \angle DOC + \angle COB = \angle EOB$$

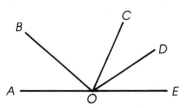

**Postulate 7.** The sum of all the adjacent angles about a point on one side of a straight line is equal to one straight angle. Thus if we measure $\angle AOB + \angle BOC + \angle COD + \angle DOE$, the total should be 180°.

---

Two angles whose sum is 90°, or one right angle, are called **complementary.** Each of the angles is called the complement of the other.

---

In the figure below, $\angle AOB$ is the complement of $\angle BOC$, or 35° is complementary to 55°, or 55° is complementary to 35°.

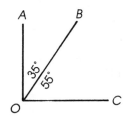

Two angles whose sum is 180°, or a straight angle, are said to be **supplementary** to one another.

---

In the figure below, $\angle AOC$ is the supplement to $\angle COB$, or 150° is supplementary to 30°, or 30° is supplementary to 150°.

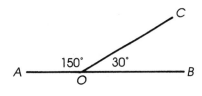

The postulates that follow concerning complementary and supplementary angles are mostly corollaries of axioms and postulates already stated.

**Postulate 8.** All right angles are equal. Since all straight angles are equal (Postulate 5) and halves of equals are equal (Axiom 5).

**Postulate 9.** When one straight line meets another, two supplementary angles are formed.

$\angle 1 + \angle 2 = \angle AOB$, which is a straight angle. (Axiom 6)

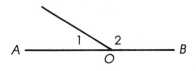

**Postulate 10.** Complements of the same angle or of equal angles are equal. (Axiom 3)

**Postulate 11.** Supplements of the same angle or of equal angles are equal. (Axiom 3)

**Postulate 12.** If two adjacent angles have their exterior sides in a straight line, they are supplementary.

**Postulate 13.** If two adjacent angles are supplementary, their exterior sides are in the same straight line.

**Vertical angles** are the pairs of opposite angles formed by the intersection of straight lines. Here $\angle 1$ and $\angle 2$ are vertical, as are $\angle 3$ and $\angle 4$.

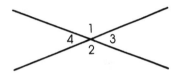

$\angle 1$ and $\angle 2$ are vertical angles. $\angle 5$ and $\angle 6$ are also vertical angles.

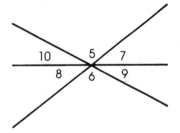

## The Method of Demonstration in Geometry

A **proposition** is a statement of either a **theorem** or a **problem.**

A theorem is a relationship to be demonstrated.

A problem is a construction to be made.

In proving theorems or the correctness of constructions, the procedure is as follows.

If the proposition is a theorem requiring proof, we break it up into its two parts: the hypothesis and the conclusion. In the hypothesis certain facts are assumed. We use these given facts in conjunction with other previously accepted geometric propositions to prove the conclusion.

If the proposition is a problem, we make the construction and then proceed to prove that it is correct. We do this by listing the given elements and bringing forward previously established geometric facts to build up the necessary proof of correctness.

For example, let us take the statement, vertical angles are equal. This theorem is given as Proposition 1 in many geometry textbooks, and is presented as follows.

*Given:* Vertical angles 1 and 2 as in the diagram.

*Prove:* $\angle 1 = \angle 2$.

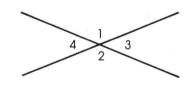

| Steps | Reasons |
|---|---|
| 1. $\angle 2$ is the supplement of $\angle 3$. | 1. Two angles are supplementary if their sum is a straight $\angle$. |
| 2. $\angle 1$ is the supplement of $\angle 3$. | 2. Same as Reason 1. |
| 3. $\angle 1 = \angle 2$. | 3. Supplements of the same $\angle$ are equal. (Postulate 11) |

## Postulates Concerning Parallels

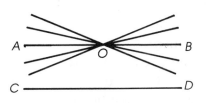

1. Through a given point only one line can be drawn parallel to a given line.
   In the diagram, the only line that can be drawn parallel to *CD* through point *O* is *AB*.

2. Two intersecting lines cannot both be parallel to a third straight line.

3. Two straight lines in the same plane, if produced, either will intersect or else are parallel.

## Definitions

A **transversal** is a line that intersects two or more other lines.

When a transversal cuts two parallel or intersecting lines, various angles are formed. The names and relative positions of these angles are important. The relationship of angles as shown in the following diagram should be memorized.

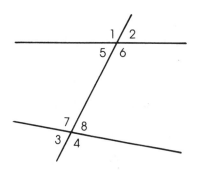

Angles 1, 2, 3, 4 are termed exterior angles.

Angles 5, 6, 7, 8 are termed interior angles.

Angles 1 and 4 ⎱ ⎰ are pairs of alternate
Angles 2 and 3 ⎰ ⎱ exterior angles.

Angles 5 and 8 ⎱ ⎰ are pairs of alternate
Angles 6 and 7 ⎰ ⎱ interior angles.

Angles 1 and 7
Angles 2 and 8 ⎰ are pairs of
Angles 5 and 3 ⎱ corresponding angles.
Angles 6 and 4

Theorem 1. If two straight lines are parallel to a third straight line, they are parallel to each other.

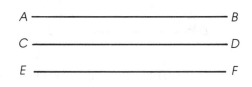

*Given:* *AB* and *EF* are parallel to *CD*.

*Prove:* *AB* is parallel to *EF*.

If *AB* is not parallel to *EF* the two lines would intersect and they would then be two intersecting lines parallel to a third straight line. But this is impossible according to Parallel Postulate 2. Hence *AB* must be parallel to *EF*.

### Relationships Formed by Parallels and a Transversal

If two parallel lines are cut by a transversal, certain definite relationships will always be found to exist among the angles that are formed by the parallel lines and the transversal.

If we take the rectangle *ABCD*, we know that the opposite sides are parallel and equal and that all the angles are right angles. If we then draw the diagonal *DB* we have formed two triangles, △*DAB* and △*DCB*.

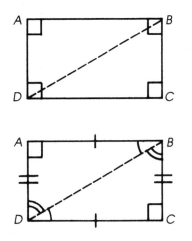

In triangles *DAB* and *DCB* we know *AD* = *CB*, *AB* = *DC* and ∠*A* = ∠*C*. As will be shown in the section on triangles, when two sides and the included angle of one triangle are equal to two sides and the included angle of another, the two triangles are said to be congruent. This means that all their corresponding sides and angles are equal. In the diagram the corresponding sides and angles of each triangle are marked with matched check marks.

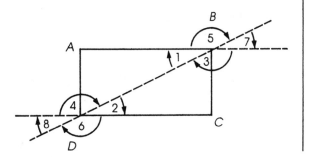

If we extend lines *AB* and *CD*, we have two parallel lines cut by a transversal. We number the related angles for convenience, and the following relationships become evident.

The angle relationships that occur when two parallel lines are cut by a transversal may be stated as follows.

**1.** The alternative interior angles are equal.

$$\angle 1 = \angle 2 \text{ and } \angle 3 = \angle 4$$

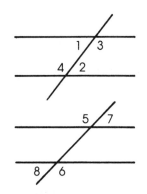

**2.** The alternative exterior angles are equal.

$$\angle 5 = \angle 6 \text{ and } \angle 7 = \angle 8$$

**3.** The corresponding angles are equal.

$$\angle 4 = \angle 5, \angle 3 = \angle 6 \text{ and}$$
$$\angle 2 = \angle 7, \angle 1 = \angle 8$$

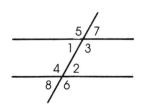

**4.** The two interior angles on the same side of a transversal are supplementary.

∠1 is supplementary to ∠4 and
∠3 is supplementary to ∠2

**5.** The two exterior angles on the same side of a transversal are supplementary.

∠5 is supplementary to ∠8 and
∠7 is supplementary to ∠6

These angle relationships may now be used to prove that certain straight lines are parallel. Such proofs are represented by the converses of statements 1 to 5, by the following theorems.

### Theorems on Parallel Lines

**Two lines are parallel if:**

**Theorem 2.** A transversal to the lines makes a pair of alternate interior angles equal.

**Theorem 3.** A transversal to the lines makes a pair of alternate exterior angles equal.

**Theorem 4.** A transversal to the lines makes a pair of corresponding angles equal.

**Theorem 5.** A transversal to the lines makes a pair of interior angles on the same side of the transversal supplementary.

**Theorem 6.** A transversal to the lines makes a pair of exterior angles on the same side of the transversal supplementary.

A corollary that follows from these theorems follows.
**Corollary 1.** If two lines are perpendicular to a third line, they are parallel.

This can be easily proved by showing alternate interior angles equal as ∠1 = ∠2, or corresponding angles equal, as ∠1 = ∠3.

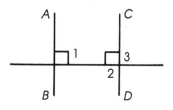

We may summarize the relationships of the angles formed by parallel lines cut by a transversal as follows:
(a) The four acute angles formed are equal.
(b) The four obtuse angles formed are equal.
(c) Any one of the acute angles is the supplement of any one of the obtuse angles; that is, their sum equals 180°.

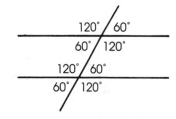

## Exercise Set 10.3

**1.** ∠1 coincides with ∠2. ∠1 = 30°. Find ∠2.

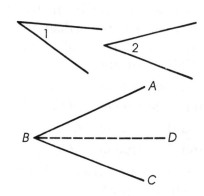

**2.** *BD* is the bisector of ∠*ABC*, which is 45°. Find ∠*ABD*.

**3.** ∠1 = ∠5, ∠2 = ∠1, and ∠3 = ∠5. What is the relationship between:

(a) ∠1 and ∠3
(b) ∠2 and ∠5
(c) ∠4 and ∠7

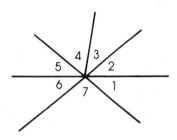

**4.** From the same figure list the pairs of adjacent angles.
**5.** From the same figure list the pairs of vertical angles.

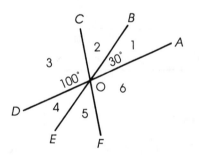

**6.** In the accompanying figure the opposite angles are vertical angles; ∠1 = 30° and ∠3 = 100°. Find the remaining four angles.
**7.** Find the value of ∠*AOC*.
**8.** Find the value of ∠*AOD*.
**9.** Find the value of ∠*BOE*.
**10.** Find the value of ∠*FOB*.
**11.** How many degrees are there in ¾ of a right triangle?

**12.** How many degrees are there in ⅔ of a right triangle?
**13.** How many degrees are there in ½ of a right triangle?
**14.** How many degrees are there in ⅓ of a right triangle?
**15.** How many degrees are there in ¼ of a right triangle?
**16.** Find the complement of 68°.
**17.** Find the complement of 45°.
**18.** Find the complement of 55°.
**19.** Find the complement of 32°.
**20.** Find the complement of 5°.
**21.** Find the complement of 33°30′.
**22.** What is the supplement of 25°?
**23.** What is the supplement of 125°?
**24.** What is the supplement of 44°?
**25.** What is the supplement of 88°?
**26.** What is the supplement of 74°30′?
**27.** What is the supplement of 78°30′?

In the following diagrams, identify the kinds of angles indicated.

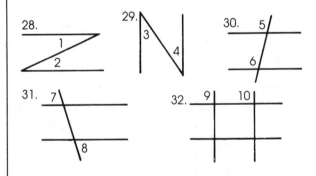

**33.** If ∠3 = 50°, what are the values of ∠1, ∠2, and ∠4?
**34.** If ∠5 = 40°, what are the values of ∠6, ∠7 and ∠8?

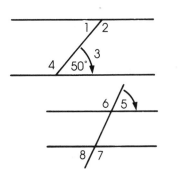

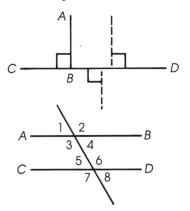

**35.** *AB* is perpendicular to *CD*. Why would any other line that makes a 90° angle with *CD* be parallel to *AB*?

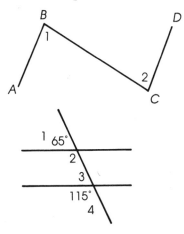

**36.** Tell why *AB* is parallel to *CD* if given:

(a) ∠3 and ∠6

(b) ∠1 and ∠5

(c) ∠2 and ∠7

**37.** *Given:* ∠1 = ∠2.

*Prove:* *AB* is parallel to *CD*.

**38.** *Given:* ∠1 = 65° and ∠4 = 115°.

*Prove:* The two horizontal lines are parallel.

**39.** If Broadway cuts across Canal Street at an angle of 70°, at what angle does it cut across Broome and Spring streets, which are parallel to Canal Street?

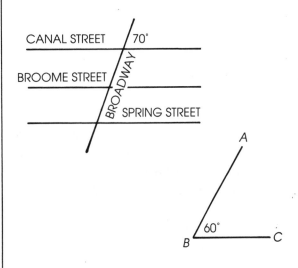

**40.** Given ∠*ABC* = 60°, construct a line parallel to *BC* using the principle of corresponding angles being equal.

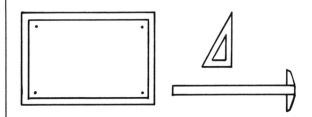

**41.** Using the drawing board, T-square, and triangle pictured, how would you construct two angles the sides of which are parallel to each other?

## 10.4 Triangles

A **triangle** is a three-sided figure, the sides of which are straight lines. If we close off any angle, a triangle is formed.

Triangles are classified according to their sides as scalene, isosceles, and equilateral.

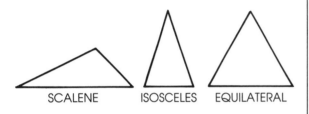

SCALENE    ISOSCELES    EQUILATERAL

A **scalene triangle** is one in which no two sides are equal. An **isosceles triangle** is one in which two sides are equal. An **equilateral triangle** is one with three sides equal.

Triangles may also be classified with respect to their angles as equiangular, right, acute, and obtuse.

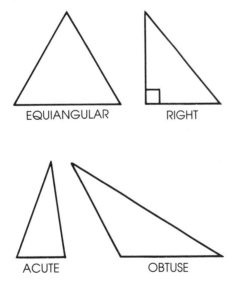

EQUIANGULAR    RIGHT

ACUTE    OBTUSE

An **equiangular triangle** is one in which all the angles are equal, or each angle measures 60°.

A **right triangle** contains one right angle, which is often indicated by placing a small square at the right angle.

An **acute triangle** is one in which all angles are less than right angles.

An **obtuse triangle** has one angle greater than a right angle.

Note that an equiangular triangle is always equilateral; a right triangle may be either scalene or isosceles; an acute triangle may be either scalene, isosceles, or equilateral; and an obtuse angle may be either scalene or isosceles.

Note also that either the scalene or the isosceles triangle may be right, acute, or obtuse. The scalene triangle cannot be equiangular, but the isosceles can, since the equilateral triangle may be considered a special type of isosceles triangle.

It is a basic theorem that the sum of the angles of any triangle is equal to 180°. The **height, or altitude, of a triangle** is the perpendicular distance from the base to the vertex of the opposite angle.

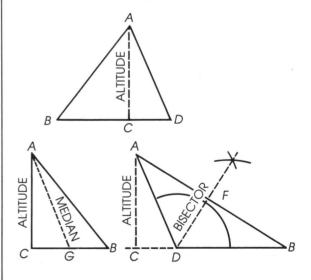

A **median** is a line drawn from any vertex of a triangle to the middle of the opposite side. See *AG* in the figure above.

The **bisector** of an angle is the line which divides it into two equal angles. *DF* bisects ∠*BDA* in the figure above.

The **perimeter** of any figure is the entire distance around the figure.

To demonstrate some fundamental relationships between lines and angles of triangles, a method of proving triangles to be congruent is used.

**Congruent figures** are those which can be made to coincide or fit on one another. Thus if two triangles coincide in all their parts, they are congruent.

The symbol for congruence is ≅.

In triangles that are congruent, the respective equal angles and equal sides that would coincide if one figure were placed on top of the other, are called corresponding angles and corresponding sides. Corresponding parts of congruent figures are equal.

In geometry, the corresponding parts of corresponding figures are frequently indicated by using corresponding check marks on the respective parts. For example, the corresponding parts in the congruent triangles below are marked with check marks of the same kind.

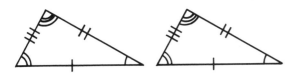

**Seven Theorems on Congruence**

**Theorem 7.** Two triangles are congruent if two sides and the included angle of one are equal respectively to two sides and the included angle of the other.

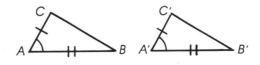

According to this theorem we are given ∠*ABC* is congruent to △*A'B'C'*, with *AC* = *A'C'*, *AB* = *A'B'*, and ∠*A* = ∠*A'*.

If we construct the figure with the given equal parts and then place △*ABC* on △*A'B'C'* so that the given equal parts correspond, it will be seen that the third line, *CB*, coincides with *C'B'*, making the triangles congruent at all points. Thus all corresponding parts not given may also be assumed to be respectively equal.

For example, construct *AC* and *A'C'* to equal ⅜ inch; ∠*A* and ∠*A'* = 60°; *AB* and *A'B'* = ¾ inch.

Then we measure the distances between *CB* and *C'B'* and find them to be equal. If we measure angles *C* and *C'* and angles *B* and *B'*, we will find these pairs to be equal as well.

Proving congruence by this theorem is known as the side angle side method. It is abbreviated SAS.

By employing a similar approach we can readily verify the following theorems on the correspondence of triangles.

**Theorem 8.** Two triangles are congruent if two angles and the included side of one are equal respectively to two angles and the included side of the other.

This is known as the angle side angle theorem, and is abbreviated ASA.

**Theorem 9.** Two triangles are congruent if the sides of one are respectively equal to the sides of the other.

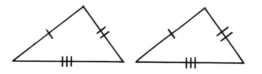

This is known as the side side side theorem, and is abbreviated SSS.

**Theorem 10.** Two triangles are congruent if a side and any two angles of one are equal to the corresponding side and two angles of the other.

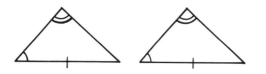

This is known as the side angle angle theorem, and is abbreviated SAA.

**Theorem 11.** Two right triangles are equal if the sides of the right angles are equal respectively.

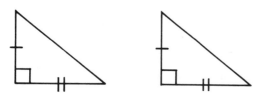

Since the included right angles are equal, this theorem is really a special case of SAS.

**Theorem 12.** Two right triangles are equal if the hypotenuse and an acute angle of one are equal to the hypotenuse and an acute angle of the other.

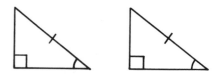

Since the right angles are equal, this theorem is a special case of SAA.

**Theorem 13.** Two right triangles are congruent if a side and an acute angle of one are equal to a side and corresponding acute angle of the other.

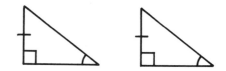

Since the right angles are equal, this is again a special case of SAA.

The general properties of the triangle not only form the foundation of trigonometry, but they also find a side application in the analysis and measurement of straight-sided plane figures of every kind.

One of the most important facts about triangles is that, regardless of the shape or size of any triangle, the sum of the three angles of a triangle is equal to a straight angle, or 180°. Presented as a theorem this proposition is easily proved.

**Theorem 14.** The sum of the angles of a triangle is equal to a straight angle.

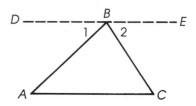

| Given: △ABC. | Prove: ∠A + ∠B + ∠C = a straight angle. |
|---|---|
| **Steps** | **Reasons** |
| 1. Through *B* draw *DE* parallel to *AC*. | 1. Parallel Postulate 1. |
| 2. ∠1 = ∠A. | 2. Alternate interior angles of parallel lines are equal. |
| 3. ∠2 = ∠C. | 3. Same as Reason 2. |
| 4. ∠1 + ∠B + ∠2 = a straight angle. | 4. By definition, since the exterior sides lie in a straight line. |
| 5. Therefore, ∠A + ∠B + ∠C = a straight angle. | 5. Substituting ∠A and ∠C for ∠1 and ∠2 in Step 4 by Axiom 7. |

From this knowledge of the sum of the angles of a triangle come the following corollaries concerning triangles.

**Corollary 1.** Each angle of an equiangular triangle is 60°.

Since the angles of an equiangular triangle are equal, each angle equals 180° ÷ 3, or 60°.

**Corollary 2.** No triangle may have more than one obtuse angle or right angle.

180° minus 90° or more leaves 90° or less, to be split between the two remaining angles, and therefore each of the two remaining angles must be acute, that is, less than 90°.

**Corollary 3.** The acute angles of a right triangle are complementary.

180° minus 90° leaves two angles whose sum equals 90°.

**Corollary 4.** If two angles of one triangle are equal respectively to two angles of another, the third angles are equal.

This truth is supported by Axiom 3, that is, if equals are subtracted from equals the remainders are equal.

**Corollary 5.** Any exterior* angle of a triangle is equal to the sum of the two remote interior angles.

*An exterior angle of a triangle is the angle formed by a side and the extension of its adjacent side. Every triangle has six exterior angles as shown in this diagram.

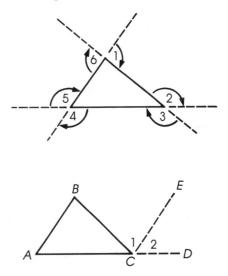

Thus in △ABC if we extend *AC* to *D* and draw *CE* parallel to *AB*, we have the two parallel lines *AB* and *CE* cut by the transversal *AD*. Therefore, ∠1 = ∠B and ∠2 = ∠A, so that ∠1 + ∠2, or ∠BCD = ∠A + ∠B.

A few characteristic properties of frequently used special triangles are worth noting at this point.

**Theorem 15.** The base angles of an isosceles triangle are equal.

By definition the sides of an isosceles triangle are equal.

Therefore, if we draw the bisector *BD* of ∠*B* it is readily seen that △*ABD* is congruent to △*CBD* by SAS. Hence ∠*A* = ∠*C*.

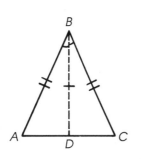

This theorem may be stated in another way, as follows:

**Theorem 16.** If two sides of a triangle are equal, the angles opposite those sides are equal.

The following corollaries may readily be seen to follow from this theorem.

**Corollary 1.** If two sides of a triangle are equal, the angles opposite these sides are equal and the triangle is isosceles.

**Corollary 2.** The bisector of the apex angle of an isosceles triangle is perpendicular to the base, bisects the base, and is the altitude of the triangle.

**Corollary 3.** An equilateral triangle is equiangular.

**Theorem 17.** If one acute angle of a right triangle is double the other, the hypotenuse is double the shorter side.

We can restate this by saying that in a 30° to 60° right triangle, the hypotenuse equals twice the shorter side.

The following properties of bisectors, altitudes, and medians of triangles are fre-

quently applied to practical design and construction problems.

**Theorem 18.** Every point in the perpendicular bisector of a line is equidistant from the ends of that line.

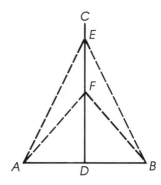

If *CD* is the perpendicular bisector of *AB*

Then *DA* = *DB*
    *FA* = *FB*

**Theorem 19.** Every point in the bisector of an angle is equidistant from the sides of the angle.

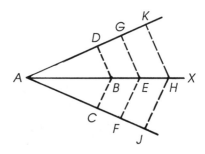

If *AX* is the bisector of ∠*A*

Then *BC* = *BD*, *EF* = *EG*, *HJ* = *HK*

**Theorem 20.** The perpendicular is the shortest line that can be drawn from a point to a given line.

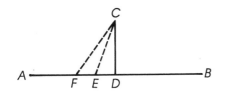

If $CD \perp AB$

Then $CD < CE$, $CD < CF$, etc.

**Theorem 21.** The three bisectors of the sides of a triangle meet in one point which is equidistant from the three vertices of the triangle.

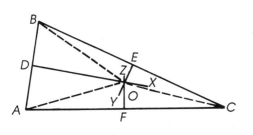

If $DX$, $EY$ and $FZ$ are bisectors of the sides $AB$, $BC$, and $CA$

Then $AO = BO = CO$, and is equal to the radius of the circle circumscribing $\triangle ABC$

Note that this fact is often used as a method for finding the center of a circular object. The procedure consists in inscribing a triangle in the circle and constructing the bisectors of the sides. The point at which they meet is the center of the circle.

**Theorem 22.** The three bisectors of the angles of a triangle meet in one point which is equidistant from the three sides of the triangle.

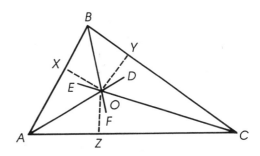

If $AD$, $BF$, and $CE$ are bisectors respectively of angles $A$, $B$, and $C$

Then $OX = OY = OZ$, and these segments are equal to the radius of a circle inscribed in $\triangle ABC$.

Note that this geometric theorem is employed as a method for determining the largest circular pattern that can be cut out of a triangular piece of material.

Try to carry out the constructions involved in the theorems of this section. Check the accuracy of the constructions by determining whether the constructed parts fit the hypothesis of the theorem. These constructions are applied daily in architecture, carpentry, art, machine work, and manufacturing.

## Exercise Set 10.4

Use a protractor, compass, and straight-edge to answer these questions.

1. Two angles of a triangle are 62° and 73°. What does the third angle equal?
2. How many degrees are there in the sum of the angles of a triangle?
3. What is the value of an exterior angle of an equilateral triangle?
4. In a certain right triangle the acute angles are $2x$ and $7x$. What is the size of each angle?

**5.** An exterior angle at the base of an isosceles triangle equals 116°. What is the value of the vertex angle?

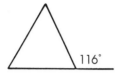

**6.** In a certain triangle one angle is twice as large as another and three times as large as the third. How many degrees are there in each angle?

**7.** Draw an equilateral triangle and by it find the ratio between the diameter of the inscribed circle and the radius of the circumscribed circle.
Hint: Refer to Theorems 20 and 22.

**8.** *Given:* ∠1 = ∠4.

*Prove:* △ABC is isosceles.

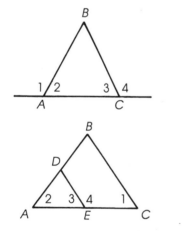

**9.** *Given:* BA = BC and DE is parallel to BC.

*Prove:* DE = DA.

Mark corresponding parts with corresponding check marks. Use the method of demonstration shown under Theorem 14, following lines of reasoning similar to that used in connection with Theorem 7.

**10.** *Given:* AB = AD and ∠1 = ∠2.

*Prove:* △ABC is congruent to △ADC.

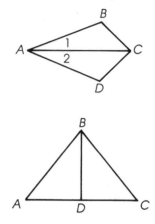

**11.** *Given:* BD is perpendicular to AC and D is the mid-point of AC.

*Prove:* △ABD is congruent to △CBD.

**12.** *Given:* ∠3 = ∠5 and AE is the bisector of BD.

*Prove:* △ABC is congruent to △EDC.

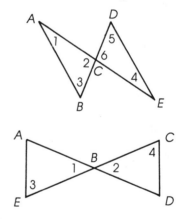

**13.** *Given:* AD and CE bisect each other.

*Prove:* AE is parallel to CD.

**14.** *Given:* AD = BC and AC = BD.

*Prove:* △BAD is congruent to △ABC and ∠1 = ∠2.

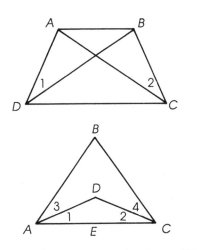

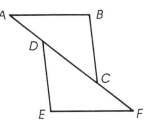

**15.** *Given: AB = CB and AD = CD.*

*Prove:* ∠1 = ∠2.

Hint: Draw *BD* and then extend it to meet *AC* at *E*.

**16.** *Given: AB = EF, AB is parallel to EF, and BC is parallel to DE.*

*Prove: BC = DE.*

## 10.5 Circles

A **circle** is a curved line with every point equally distant from the center, a point within the circle. A **radius** of a circle is a line drawn from the center to any point on the circle.

The **diameter** of a circle is a straight line drawn from any point on the circle, through the center of the circle to the opposite side of the circle. The diameter is twice the length of the radius. A **chord** is a segment whose endpoints lie on the circle.

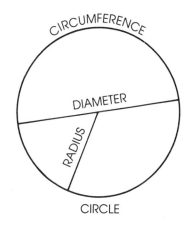

The **circumference** of a circle is the distance around the circle. The circumference can also be described as the perimeter of the circle.

**Pi,** written **π** is the name given to the ratio expressed by dividing the circumference of any circle by its diameter. It is a constant approximately equal to 3.14. If we measure the distance around any circle, measure its diameter, and then divide the distance around the circle by the diameter, our result should be approximately 3.14.

---

**Formula for Pi**

$\pi = \dfrac{C}{d}$, where $C$ = circumference and $d$ = diameter;

or $\pi = \dfrac{C}{2r}$, where $r$ = radius.

---

To find the circumference of a circle, multiply the diameter by pi.

---

**Formula for the Circumference of a Circle**

$$C = \pi d; \text{ or } C = 2\pi r$$

---

EXAMPLE 1: The spoke of a wheel is 21 inches long. Find the circumference of the wheel.

SOLUTION: We know the formula for the circumference of a circle is $C = 2\pi r$. We use 3.14 as an approximation for pi.

$$C = 2\pi r$$
$$C = 2 \times 3.14 \times 21 \text{ inches}$$
$$C = 131.88 \text{ inches}$$

The circumference of the wheel is 131.88 inches.

EXAMPLE 2: The circumference of a pulley is 33 inches. What is its diameter?

SOLUTION: We solve for $d$, or diameter, using the formula for circumference.

$$C = 2\pi r$$
$$\frac{c}{\pi} = d$$
$$d = 33 \text{ inches}/3.14$$
$$= 10.51 \text{ inches}$$

The diameter of the pulley is approximately $10\frac{1}{2}$ inches long.

## Exercise Set 10.5

1. The circumference of a wheel is 110 inches. How long is one of its spokes to the nearest inch?
2. To make a circular coil for a magnet, we need 49 loops of wire. How much wire do we need if the diameter of the coil is 4 inches?

3. A round window has a 12-inch radius. What is the circumference of the window?
4. If a lampshade has a 7-inch radius, what is its circumference?
5. If a circle has a 25-foot radius, what is its circumference?

# 10.6 Quadrilaterals and Other Polygons

A **polygon** is a plane geometric figure bounded by three or more line segments that join without crossing one another. For instance, any triangle is a polygon.

The **vertices** of a polygon are the points where two sides meet.

A **diagonal** of a polygon joins two nonconsecutive vertices. How many diagonals does a triangle have? None. How many diagonals can a four-sided figure have? Two.

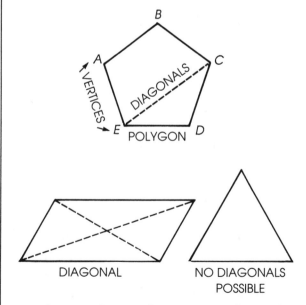

Polygons derive their names from the number of sides and the types of angles included.

**Quadrilaterals** are polygons with four sides.

There are several types of quadrilaterals, and in this section we will define the rectangle, the square, the rhombus, the trapezoid, the parallelogram, and the trapezium.

A **parallelogram** is a quadrilateral where both pairs of opposite sides are parallel. Both pairs of opposite angles are equal, and both pairs of opposite sides are equal in length.

A **square** is a parallelogram with four right angles and four sides equal in length.

A **rectangle** is a parallelogram with four right angles. Each pair of opposite sides is equal in length.

A **rhombus** is a parallelogram with two adjacent sides of equal length.

A **trapezoid** is a quadrilateral having one pair of parallel sides.

A **trapezium** is a quadrilateral with no two sides parallel.

The **height,** or **altitude, of a parallelogram** is the distance perpendicular from the base to the opposite side.

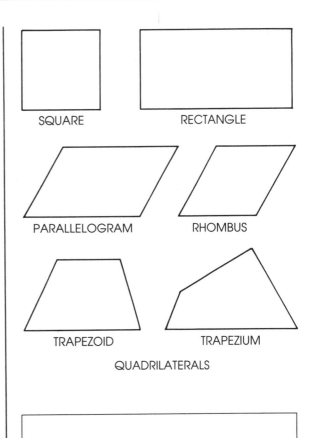

SQUARE    RECTANGLE

PARALLELOGRAM    RHOMBUS

TRAPEZOID    TRAPEZIUM

QUADRILATERALS

The perimeter of a quadrilateral is the sum of the length of each of the four sides.

**PRACTICALLY SPEAKING 10.6**

Dagmar is planning an herbal garden for her backyard. The garden will be rectangular, measuring 12 feet long by 8 feet wide. She plans on putting white stones that are 6 inches long around the perimeter.

1. What is the perimeter of the garden?
2. How many stones will Dagmar need?

See Appendix F for the answers.

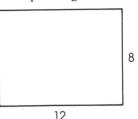

## Exercise Set 10.6

Name each of the following figures.

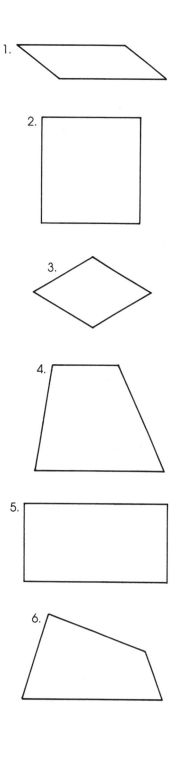

1.

2.

3.

4.

5.

6.

Solve each of the following problems.

**7.** If a parallelogram has one side 4 feet long, and an adjacent side 7 feet long, what is the perimeter of the parallelogram?

**8.** If a rhombus has sides 6 feet long, what is its perimeter?

**9.** If a square has sides 7 feet long, what is the perimeter of the square?

**10.** The foundation of a house is rectangular in shape. If the front is 550 feet long, and the sides are 390 feet long, what is the distance all the way around the foundation?

## 10.7 Similar Plane Figures

In ordinary language plane figures are similar when they are alike in all respects except size. For instance, all circles are obviously similar.

Two polygons are **similar** when the angles of one are respectively equal to the angles of the other in the same consecutive order.

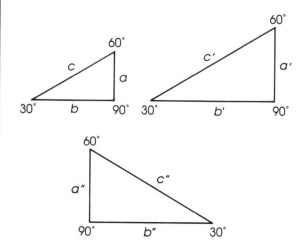

If the consecutive order of the angles is the same, it makes no difference if they follow each other clockwise in one figure and counterclockwise in the other. Such figures will still be similar because either may be considered as having been reversed like an image in a mirror.

In the case of triangles it is impossible not to arrange the angles in the same consecutive order, so that two triangles are similar if only their angles are equal.

In the preceding diagram all three triangles are similar because they all have the same angles.

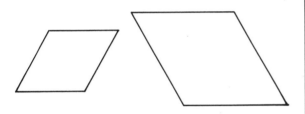

The two rhombuses are similar, even though the direction of the lines in one is the reverse of that in the other.

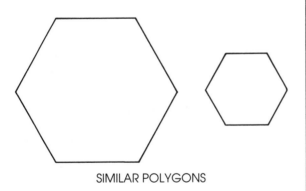

SIMILAR POLYGONS

If two polygons are similar, the ratio of any line in one polygon to the corresponding line in the other polygon applies to all the lines that correspond in the two figures.

If two polygons are similar, the ratio of their areas is that of the squares of corresponding lines.

These rules apply to drawings, photographs, and complex blueprints, as well as to simple geometric figures. Because of the broad applicability of the rules governing similar polygons, we have generalized the whole subject here.

To find the length of any line in a plane figure that is similar to another plane figure, apply the ratio that exists between any other two corresponding lines.

EXAMPLE 1: In a rhombus measuring 4 inches on a side the longer diagonal is $5\frac{1}{2}$ inches. How long would this diagonal be in a similar rhombus measuring 7 inches on a side?

SOLUTION:

$$D : d :: S : s$$

$$\frac{D}{5\frac{1}{2}} = \frac{7}{4}$$

$$D = \frac{7 \times 5\frac{1}{2}}{4}$$

$$= \frac{38\frac{1}{2}}{4}$$

$$= \frac{77}{2 \times 4}$$

$$= 9\frac{5}{8} \text{ inches}$$

To find the area of a plane figure that is similar to another plane figure with a known area, determine the ratio of any two corresponding lines in the two figures and make the required area proportional to the squares of these lines.

EXAMPLE 2: A trapezium in which one of the sides measures 6 inches has an area of 54 square inches. What is the area of a similar trapezium where a corresponding side measured 15 inches?

SOLUTION:

$$A' : A :: S^2 : s^2$$

$$\frac{A'}{54} = \frac{15^2}{6^2}$$

$$A' = \frac{225 \times 54}{36}$$

$$= \frac{225 \times 3}{2}$$

$$= 337\frac{1}{2} \text{ inches}$$

## Exercise Set 10.7

Find out whether or not each pair of figures is similar.

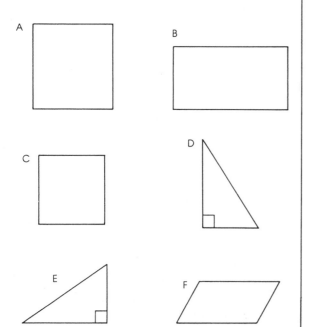

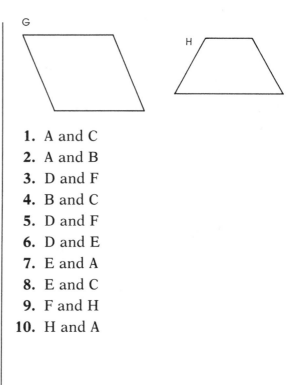

1. A and C
2. A and B
3. D and F
4. B and C
5. D and F
6. D and E
7. E and A
8. E and C
9. F and H
10. H and A

## Chapter 10 Glossary

**Acute Angle**   Any angle that measures less than 90°.

**Acute Triangle**   A triangle with all angles less than 90°.

**Adjacent Angles**   Angles that have a common vertex and a common side between them.

**Angle**   The figure formed by two lines meeting at a common point.

**Axiom**   Basic mathematical principle.

**Bisector**   The line that divides an angle into two equal pieces.

**Chord**   A segment whose endpoints lie on the circle.

**Circle**   A curved line with every point equally distant from the center.

**Circumference**   The distance around the circle.

**Complementary Angles**   Two angles whose sum is 90°.

**Congruent Figures** Figures that coincide, or fit, on one another.

**Corollary** A geometric truth that follows from one previously given and needs little or no proof.

**Degree** The unit used to measure angles.

**Diagonal of a Polygon** A line segment joining two nonconsecutive vertices.

**Diameter** A straight line drawn from any point on the circle through the center, to the opposite side of the circle.

**Equiangular Triangle** A triangle with three equal angles, each measuring 60°.

**Equilateral Triangle** A triangle with three equal sides.

**Geometric Figure** A point, line, plane, solid, or a combination of these.

**Geometry** The branch of mathematics that deals with spatial relationships.

**Height of a Parallelogram** The height of a parallelogram is the distance perpendicular from the base to the opposite side.

**Height of a Triangle** The height of a triangle is the perpendicular distance from the base to the vertex of the opposite angle.

**Hexagon** A regular six-sided figure.

**Horizontal Line** A straight line that goes straight across from left to right.

**Hypotenuse** The side opposite the right angle in a right triangle.

**Isosceles Triangle** A triangle with two equal sides.

**Line** The intersection of two surfaces. It has length, but does not have width or thickness.

**Median** A line drawn from any vertex of a triangle to the middle of the opposite side.

**Obtuse Angle** Any angle that measures more than 90°, but less than 180°.

**Obtuse Triangle** A triangle with one angle greater than 90°.

**Parallel Lines** Two lines that are equally distant from one another at all points.

**Parallelogram** A quadrilateral where both pairs of opposite sides are parallel.

**Perimeter** The entire distance around any figure.

**Perpendicular Lines** Two lines that form right angles when they intersect.

**Pi** The ratio expressed by dividing the circumference of any circle by its diameter.

**Point** The position of the intersection of two lines. It does not have length, width, or thickness.

**Polygon** A plane geometric figure bounded by three or more line segments that join without crossing one another.

**Problem** A construction to be made.

**Proposition** The statement of either a theorem or a problem.

**Quadrilateral** A polygon with four sides.

**Radius** A line drawn from the center of a circle to any point on the circle.

**Rectangle** A parallelogram with four right angles.

**Rhombus** a parallelogram with two adjacent sides of equal length.

**Right Angle** An angle that measures 90°.

**Right Triangle** A triangle that contains one right angle.

**Scalene Triangle** A triangle with no equal sides.

**Sides** The lines that form an angle.

**Similar Polygons** Two polygons with equal corresponding angles, and where the ratio of any line in one polygon to the corresponding line in the other polygon is the same.

**Square** A parallelogram with four right angles and four sides equal in length.

**Straight Angle** An angle that measures 180°.

**Supplementary Angles** Two angles whose sum is 180°.

**Theorem** A relationship to be demonstrated.

**Transversal** A line that intersects two or more other lines.

**Trapezium** A quadrilateral with no two sides parallel.

**Trapezoid** A quadrilateral having one pair of parallel sides.

**Triangle** A three-sided figure, the sides of which are straight lines.

**Vertex** A point where two lines meet.

**Vertical Angles** The pairs of opposite angles formed by the intersection of straight lines.

**Vertical Line** A straight line that goes straight up and down.

## Chapter 10 Test

For each problem, five answers are given. Only one answer is correct. After you solve each problem, check the answer that agrees with your solution.

1. A triangle with two equal sides is called

   **A)** isosceles     **D)** acute
   **B)** equilateral    **E)** obtuse
   **C)** right

2. An angle that measures more than 90° is

   **A)** vertical     **D)** acute
   **B)** oblique     **E)** obtuse
   **C)** right

3. An angle that measures 90° is

   **A)** vertical     **D)** acute
   **B)** oblique     **E)** obtuse
   **C)** right

4. Angles that have a common vertex and a common side are

   **A)** horizontal    **D)** supplementary
   **B)** complementary **E)** adjacent
   **C)** oblique

5. Two angles whose sum adds up to 90° are

   **A)** horizontal    **D)** supplementary
   **B)** complementary **E)** adjacent
   **C)** oblique

6. In this figure, one pair of alternate exterior angles is

   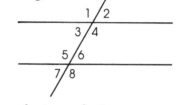

   **A)** angle 1, angle 2
   **B)** angle 5, angle 6
   **C)** angle 1, angle 8
   **D)** angle 5, angle 4
   **E)** angle 1, angle 4

7. A five-sided polygon is called a(n)

   **A)** triangle     **D)** pentagon
   **B)** parallelogram **E)** rhombus
   **C)** octogon

8. A four-sided polygon with four sides of equal length, and four equal angles is a

   **A)** square     **D)** pentagon
   **B)** parallelogram **E)** rhombus
   **C)** trapezoid

9. A triangle where no two sides are equal is

   **A)** isosceles    **D)** acute
   **B)** scalene     **E)** obtuse
   **C)** right

10. The side of a right triangle opposite the right angle is called the

    **A)** hypotenuse     **D)** median
    **B)** perimeter     **E)** altitude
    **C)** bisector

11. The distance around a circle is called the

    **A)** diameter     **D)** pi
    **B)** center     **E)** circumference
    **C)** radius

12. When the angles of one of two respective polygons are equal to the angles of the second in the same consecutive order, the polygons are

    **A)** similar     **D)** lateral
    **B)** diagonal     **E)** solid
    **C)** obtuse

13. What is the perimeter of a rectangle with a 7-inch width and a 16-inch length?

    **A)** 32 inches     **D)** 46 inches
    **B)** 63 inches     **E)** 54 inches
    **C)** 23 inches

14. What is the perimeter of an isosceles triangle with two 6-inch sides, and a base measuring 4 inches?

    **A)** 16 inches     **D)** 12 inches
    **B)** 25 inches     **E)** 24 inches
    **C)** 10 inches

15. What is the perimeter of a 12 × 14 inch-picture frame?

    **A)** 48 inches     **D)** 48 inches
    **B)** 52 inches     **E)** 56 inches
    **C)** 64 inches

16. What is the perimeter of a rhombus with 11-inch sides?

    **A)** 64 inches     **D)** 74 inches
    **B)** 19 inches     **E)** 44 inches
    **C)** 154 inches

17. If one angle measures 75°, what must a second angle measure if the two angles are to be supplementary angles?

    **A)** 35°     **D)** 75°
    **B)** 115°     **E)** 105°
    **C)** 15°

18. If an equilateral triangle has 9-inch sides, what is the perimeter of the triangle?

    **A)** 36 inches     **D)** 48 inches
    **B)** 18 inches     **E)** 24 inches
    **C)** 27 inches

19. If one of two complementary angles measures 26°, what does the other angle measure?

    **A)** 64°     **D)** 74°
    **B)** 19°     **E)** 44°
    **C)** 154°

20. The diameter of an automobile tire is 28 inches. What is its circumference?

    **A)** 66 inches     **D)** 99 inches
    **B)** 77 inches     **E)** 121 inches
    **C)** 88 inches

21. Find the radius of a circle with a 314-yard circumference.

    **A)** 10 yards     **D)** 24 yards
    **B)** 22 yards     **E)** 100 yards
    **C)** 50 yards

# Measurement of Geometric Figures

## 11.1 Area of Quadrilaterals

The **area** of a rectangular surface is the number of square units which it contains.

In finding an area the unit of measure is a square each side of which is a unit of the same type of measure as the given dimensions. To find the area of a rectangular surface 8 feet long and 5 feet wide, the measuring unit will be 1 square foot, since the length and width are measured in feet.

> To find the area of a rectangular surface multiply the two dimensions.

EXAMPLE 1: How many square yards are in a sidewalk 48 feet long and 11 feet 4 inches wide?

SOLUTION: Taking 1 square foot as the unit, a sidewalk 48 feet long and 1 foot wide will contain 48 square feet. A sidewalk of equal length and $11\frac{1}{3}$ feet wide must contain $11\frac{1}{3}$ times 48 square feet or 544 square feet. This reduces to $60\frac{4}{9}$ square yards.

11 feet 4 inches $= 11\frac{1}{3}$ feet

48 square feet $\times$ $11\frac{1}{3} = 544$ square feet

$544 \div 9 = 60\frac{4}{9}$ square yards

To find the length or the width of a rectangular surface when one dimension and the area are given, divide the area by the given dimension.

EXAMPLE 2: If a rectangular field with a 30-acre area is 1,200 yards long, how wide is the field?

SOLUTION:

> 30 acres = 145,200 square yards
> 145,200 square yards ÷ 1,200 yards = 121 yards

The field is 121 yards wide.

The area of a rectangle equals the base multiplied by the height.

$A = bh$

EXAMPLE 3: Find the area of a rectangle that is 3 inches high with a 4-inch base.

SOLUTION: $A = bh$

$$A = 4 \times 3 = 12$$

The area is 12 square inches.

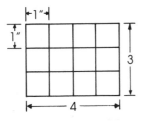

EXAMPLE 4: Find the height of a rectangle with a 16-foot base and an area of 80 square feet.

SOLUTION: $A = bh$

$$h = \frac{A}{b}$$

$$h = \frac{80}{16}$$

$$= 5 \text{ feet}$$

The area of a square is equal to the square of one of its sides.

$A = S^2$

EXAMPLE 5: Find the length of the side of a square whose area is 121 square inches.

SOLUTION: $A = S^2$

$$\sqrt{A} = S$$

$$A = \sqrt{121}$$

$$= 11 \text{ feet}$$

The perimeter of a square is equal to four times the square root of the area.

$P = 4\sqrt{A}$, or $P = 4S$

EXAMPLE 6: Find the perimeter of a square whose area is 144 square inches.

SOLUTION: $P = 4\sqrt{A}$

$$P = 4 \times 12$$

$$= 48 \text{ inches}$$

The diagonal of a square equals the square root of twice the area.

$$D = \sqrt{2A}$$

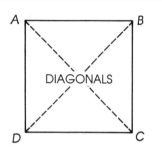

DIAGONALS

We can use the right triangle formula $c^2 = a^2 + b^2$, where $c$ represents the diagonal or hypotenuse while $a$ and $b$ are the sides to find the length of the hypotenuse.

The square of the hypotenuse of a right triangle is equal to the sum of the squares of the two other sides.

$$c^2 = a^2 + b^2$$

Any parallelogram can be converted to a rectangle without changing its area. This is shown in the following diagram.

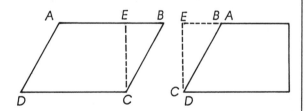

By taking $\Delta EBC$ from the figure at the left and changing its position as shown in the figure at the right, we create a rectangle without adding to or deducting from the total area.

The area of a parallelogram is equal to the product of the base times the height.

$$A = bh$$

EXAMPLE 7: Find the area of a rhombus whose base is 12 inches and whose height is 8 inches.

SOLUTION: $A = bh$

$$A = 12 \times 8$$

$$= 96 \text{ square inches}$$

The area of a trapezoid equals half the sum of the parallel sides multiplied by the height.

$$A = \frac{1}{2}h(B + b)$$

where $h$ is the perpendicular height, with parallel bases $B$ and $b$.

Make a rectangle of the trapezoid $ABCF$ by drawing a line $GH$ parallel to the two parallel sides and midway between them. The length of this line is the average of the two parallel sides $AB$ and $FC$. Perpendiculars from the midline $GH$ to the larger base $FC$ cut off triangles that are exactly equal to the triangles needed above the midline to form a rectangle of the new figure.

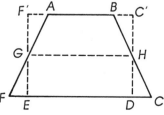

EXAMPLE 8: Find the area of a trapezoid whose bases are equal to 20 inches and 30 inches and whose height is 15 inches.

SOLUTION: $A = \dfrac{B + b}{2} \times h$

$$A = \dfrac{30 + 20}{2} \times 15$$

$$= 25 \times 15$$

$$= 375 \text{ square inches}$$

---

**PRACTICALLY SPEAKING 11.1**

George is trying to figure out how much carpeting he needs for his apartment. His bedroom measures 12 feet × 15 feet, his kitchen measures 8 feet × 8 feet, and his living room measures 15 feet × 20 feet. George does not want carpet in his bathroom.

1. How many square feet of carpeting does George need?

(See Appendix F for the answer.)

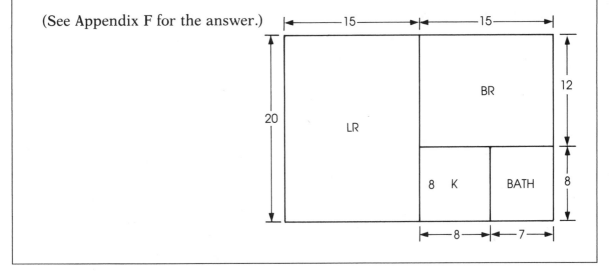

---

## Exercise Set 11.1

1. If a postage stamp measures 3 centimeters by 7 centimeters, what is the area of the stamp?

2. If a billboard is 70 feet long and 50 feet wide, what is the area of the billboard?

3. A square table has sides 3 feet long. What is the area of the table top?

4. A rectangular hangar is being built to house an airplane. What must the area of the hangar be if the plane is 110 feet long by 64 feet wide, and there must be a 20-foot allowance on all sides of the airplane?

5. How much would it cost to fertilize a square garden plot with 75-meter sides, if fertilizer costs 20 cents a square meter?

6. A square terrace with a 1,024-square-foot area is to be completely covered with flagstones. Each flagstone is 4-feet square. How many flagstones are needed to cover the terrace?

7. How much barbed wire is needed to cross diagonally over a field that is 66 feet wide by 88 feet long?

8. If we have a square wooden frame that has a 288-square-foot area, how long a piece of wood do we need to make a diagonal brace from one corner to the opposite corner of the frame?

9. If molding costs 60 cents a foot, how much would it cost to put a border of molding around a square window with an area of 81 square feet?

10. What is the area of the figure shown below?

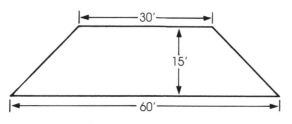

11. What is the area of the figure shown below?

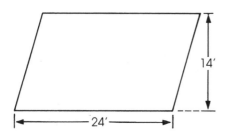

12. What is the area of a rectangle 15 feet long and 8 feet wide?

## 11.2 Area of Triangles and Circles

**Plane geometry** deals with surfaces, or with figures having two dimensions, that is, length or width.

> The area of a triangle equals $\frac{1}{2}$ the product of the base and the height.
>
> $A = \frac{1}{2}bh$

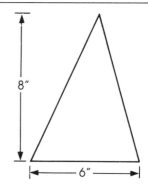

EXAMPLE 1: Find the area of the triangle shown above.

SOLUTION:     $A = \dfrac{bh}{2}$

$$= \frac{6 \times 8}{2}$$

$$= 24 \text{ square inches}$$

EXAMPLE 2: What is the height of a triangle if its area is 1 square foot and its base is 16 inches?

SOLUTION:

$A = \dfrac{bh}{2}$, therefore

$h = \dfrac{2A}{b}$

$= \dfrac{2 \times 144}{16}$

$= 18 \text{ inches}$

## Facts About Right Triangles

The **hypotenuse** of a right triangle is the side opposite the right angle.

In the figure below it is shown that the square drawn on the hypotenuse of a right triangle is equal in area to the sum of the areas of the squares drawn on the other two sides.

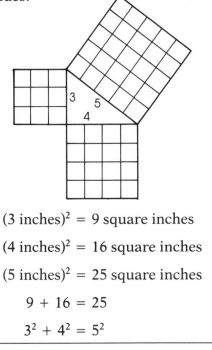

(3 inches)$^2$ = 9 square inches

(4 inches)$^2$ = 16 square inches

(5 inches)$^2$ = 25 square inches

$$9 + 16 = 25$$

$$3^2 + 4^2 = 5^2$$

> The square of the hypotenuse of a right triangle is equal to the sum of the squares of the two other sides.

Several self-evident formulas with reference to the right triangle arise from this rule.

Let $c$ = hypotenuse

$a$ = altitude

$b$ = base

**Formula 1:** $c^2 = a^2 + b^2$

**Formula 2:** $c = \sqrt{a^2 + b^2}$ (Take the square root of both sides of the first equation.)

**Formula 3:** $a^2 = c^2 - b^2$, or $b^2 = c^2 - a^2$

EXAMPLE 3: Find the hypotenuse of a right triangle whose base is 18 inches and altitude 26 inches.

SOLUTION:

$$c = \sqrt{a^2 + b^2}$$

$$= \sqrt{(18)^2 + (26)^2} \qquad \text{Substitute.}$$

$$= \sqrt{324 + 676} \qquad \text{Square.}$$

$$= \sqrt{1,000} \qquad \text{Add.}$$

$$= 31.62$$

The area of a circle equals ½ the product of the circumference and the radius.

This can be reasoned informally as follows. Any circle can be cut to form many narrow triangles as shown below. The altitude of each triangle would be equal to a radius $r$. The base would be a part of the circumference $C$. We know the area of each triangle to be equal to ½ the base times the altitude. Since $r$ is the altitude, and the sum of the bases equals the circumference, the area $= \frac{1}{2}r \times C$. Since $C = 2\pi r$, $A = \frac{1}{2}r \times 2\pi r$. Therefore, $A = \pi r^2$.

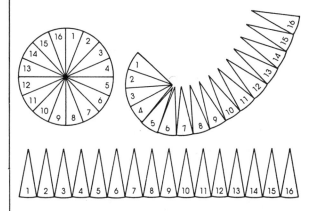

### Area of a Circle

$$A = \pi r^2$$

The area of a circle is pi times the radius squared.

EXAMPLE 4: Find the area of a circle that has a 6-inch radius.

SOLUTION:

$$A = \pi r^2$$
$$= 3.14 \times (6)^2$$
$$= 113.04 \text{ square inches}$$

EXAMPLE 5: The area of a circle is 396 square inches. Find its radius.

SOLUTION: $A = \pi r^2$

$$\frac{A}{\pi} = r^2$$

$$\sqrt{\frac{A}{\pi}} = r$$

$$r = \sqrt{\frac{396}{3.14}}$$

$$= \sqrt{126.11}$$

$$= 11.23$$

### Area of a Circular Ring

$$A = \pi R^2 - \pi r^2$$

where $R$ is the radius of the larger circle and $r$ is the radius of the smaller circle.

The area of a circular ring equals the area of the outside circle, minus the area of the inside circle.

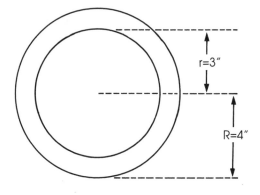

EXAMPLE: In a circular ring the outside diameter is 8 inches and the inside diameter is 6 inches. What is the area of a cross section of the ring?

SOLUTION:

$$A = \pi R^2 - \pi r^2$$
$$\text{Since } D = 8, R = 4$$
$$\text{Since } d = 6, r = 3$$

$$A = \pi(4^2 - 3^2)$$
$$= 3.14(4^3 - 3^2)$$
$$= 3.14(16 - 9)$$
$$= 3.14(7)$$
$$= 21.98$$

The area of the cross section of the ring is 21.98 square inches.

## Exercise Set 11.2

1. A field is circular, with a 96-foot diameter. Concentric with the field is a circular track with a 128-foot diam-

eter. How much will it cost to put gravel on the track, if gravel costs $1.00 per square foot?

2. The area of a cap needed to cover a round can is 50.24 inches. What is the diameter of the cap?

3. If the radius of a circle is twice as great as the radius of smaller circle, how many times as large will the area of the greater circle be than the area of the smaller circle?

4. We have four circular garden plots, each having a 14-foot radius. What must the radius be of one large circular plot with the same amount of area as the other four plots combined?

5. How much will it cost to resurface the bottom of a circular swimming pool with a 56-foot diameter, if surfacing costs 25 cents a square foot?

6. If a pizza has a 14-inch radius, how many square inches of pizza are there?

7. A triangular courtyard has a 14-foot base and a 21-foot height. What is the area of the courtyard?

8. A right triangle has one side 8 yards long, and another side 15 yards long. What is the length of the hypotenuse of the right triangle?

9. A triangular piece of cloth has a base 7 inches long and a height of 14 inches. What is the area of the cloth?

10. A right triangle has a 5-foot hypotenuse, and one side 3 feet long. What is the length of the other side?

# 11.3 Volume and Surface Area of Solids

**Solid geometry** deals with solids or bodies having three dimensions, that is, length, width, and height.

The **volume** of a rectangular solid is the number of cubic units which it contains. Another term used is capacity.

The unit of cubical measure is a cube each edge of which is a unit of the same denominator as the three given dimensions. To find the volume of a rectangular solid which is 6 feet long, 4 feet wide, and 3 feet deep, the measuring unit will be 1 cubic foot, since the type of measure of the length, width, and depth is feet.

A **rectangular solid** is one in which all the faces are rectangles. The **cube** is a special type of rectangular solid in which all the faces are equal.

> To find the **volume** of a rectangular solid, multiply together the three dimensions.

EXAMPLE 1: What is the volume of a box 3 feet 8 inches long, 3 feet 6 inches wide, and 2 feet 3 inches high?

SOLUTION: If we imagine cubes measuring 1 foot in each direction, we find that row of $3\frac{2}{3}$ cubes could be laid along the length of the box. The $3\frac{1}{2}$ feet of width would accommodate $3\frac{1}{2}$ cubes. Since the depth of the box is $2\frac{1}{4}$ feet, we would have room for $2\frac{1}{4}$ cubes. To find the volume, we multiply the three dimensions together. Our answer will be in cubic feet, because our cubes all measured 1 foot in each dimension. The volume is $3\frac{2}{3} \times 3\frac{1}{2} \times 2\frac{1}{4}$, or $28\frac{7}{8}$ cubic feet.

$$3\frac{2}{3} \text{ feet} \times 3\frac{1}{2} \text{ feet} \times 2\frac{1}{4} \text{ feet}$$
$$= 28\frac{7}{8} \text{ cubic feet}$$

To find a third dimension of a rectangular solid when the other two dimensions and the volume are given, divide the volume by the product of the two given dimensions.

EXAMPLE 2: A block of stone 7 feet long and 5 feet wide has a volume of 140 cubic feet. What is its height?

SOLUTION: A block 7 feet long, 5 feet wide, and 1 foot high would contain 35 cubic feet. To contain 140 cubic feet, a block of this length and width would have to be as many times 1 foot high as 35 is contained in 140.

$$140 \div (7 \times 5) = 4 \text{ feet}$$

To find the area of the faces of a rectangular solid, add the areas of the three pairs of opposite faces.

EXAMPLE 3: A rectangular solid measures 6 inches × 4 inches × 3 inches. What is the total area of its faces?

SOLUTION: It has two faces measuring 6 inches × 4 inches, two measuring 6 inches × 3 inches, and two measuring 4 inches × 3 inches.

$$2 (6 \times 4) + 2 (6 \times 3) + 2 (4 \times 3)$$
$$= 108 \text{ square inches}$$

To find the area of the faces of a cube, multiply the area of one face by 6.

With solids other than rectangular ones we consider the surfaces and areas of the sides as distinct from those of the bottom and top, if any exist. We call the area of the sides the **lateral area** and speak of the top as well as the bottom as **bases.**

To find the lateral surface area of a prism, multiply the perimeter of one of the bases by the height.

EXAMPLE 4: A prism 6 inches high has as its base an equilateral triangle measuring $1\frac{1}{2}$ inches on a side. What is its lateral area?

SOLUTION: $(1\frac{1}{2} + 1\frac{1}{2} + 1\frac{1}{2}) \times 6 = 27$ square inches

To find the volume of a prism, multiply the area of one of the bases by the height.

EXAMPLE 5: What is the volume of a prism 8 inches high if the area of one of the bases is $3\frac{3}{4}$ square inches?

SOLUTION: $3\frac{3}{4} \times 8 = 30$ cubic inches

To find the lateral surface area of a cylinder, multiply the circumference of one of the bases by the height.

EXAMPLE 6: What is the lateral surface area of a cylinder with a base 6 inches in diameter if its height is 7 inches?

SOLUTION: $6 \times 3.14 \times 7 = 131.88$ square inches

---

To find the volume of a cylinder, multiply the area of one of the bases by the height.

---

EXAMPLE 7: What is the volume of the cylinder in the preceding example?

SOLUTION:

$3^2 \times 3.14 \times 7 = 197.82$ cubic inches

---

To find the lateral surface area of a pyramid, multiply its slant height by the perimeter and divide by 2.

---

EXAMPLE 8: What is the lateral surface area of a triangular pyramid having a base measuring 2 inches on a side and a slant height of 9 inches?

SOLUTION:

$(2 + 2 + 2) \times 9 \div 2 = 27$ square inches

---

To find the volume of a pyramid, multiply the area of the base by the altitude not the slant height, and divide by 3.

---

EXAMPLE 9: A square pyramid 10 inches high has a base measuring 4 inches on a side. What is its volume?

SOLUTION:

$$\frac{4 \times 4 \times 10}{3} = \frac{160}{3} = 53\frac{1}{3} \text{ cubic inches}$$

---

To find the lateral surface area of a cone, multiply its slant height by the circumference of the base and divide by 2.

---

To find the volume of a cone, multiply the area of the base by the altitude and divide by 3.

---

To find the area of the surface of a sphere, multiply the square of the radius by $4\pi$.

---

EXAMPLE 10: What is the surface area of a sphere 1 foot in diameter?

SOLUTION:

$$6^2 \times 4\pi = 36 \times 4 \times 3.14$$
$$= 452.16 \text{ square inches}$$

---

To find the volume of a sphere, multiply the cube of the radius by $\frac{4\pi}{3}$.

---

EXAMPLE 11: What is the volume of a sphere 1 foot in diameter?

SOLUTION: $6^3 \times \dfrac{4\pi}{3} = 904.32$ cubic inches

## Exercise Set 11.3

1. How many cubic inches in 13 cubic feet?
2. How many cubic inches in 11 cubic yards?
3. How many cubic inches in a crate that holds 5 cubic yards, 6 cubic feet?
4. How many cubic feet in 135 cubic inches?
5. How many cubic feet in 15,552 cubic inches?
6. How many cubic yards in 466,560 cubic inches?
7. How many cubic yards in 11,106 cubic feet?
8. How many cubic inches in 1 cubic yard?
9. How many cubic inches in a block 1 foot × 2 feet × 3 feet?
10. What is the volume of a box measuring 11 inches × 6 inches × 4½ inches?
11. How many square inches of tin are needed for the top of a can that is 14 inches in diameter?
12. What is the volume of a sphere with a 6-foot diameter?
13. What is the surface area of a circular ball with an 8-inch radius?
14. A pyramid has a triangular base with a 14-square-inch area. If the height of the pyramid is 8 inches, what is the volume of the pyramid?
15. A cone-shaped building has a 200.96-square-yard base. If the height of the building is 6 yards, what is the volume?

## 11.4 Metric System

Since the **metric system** is based on decimal values, just like United States money, all ordinary arithmetical operations may be performed by simply moving the decimal point.

Consider the quantity 4.567 meters. It is made up of the following units:

> 4 meters
> 5 decimeters
> 6 centimeters
> 7 millimeters

We may read this quantity as 4.567 meters, or as 45.67 decimeters, or as 456.7 centimeters, or as 4,567 millimeters.

When it comes to dealing with square and cubic measures we must keep in mind that instead of moving the decimal point one place for each successive change in unit value, we move it two places in the case of square measurements and three places in the case of cubic measurements.

Thus, take 42.365783 square meters. This consists of 42 square meters, 36 square decimeters, 57 square centimeters, 83 square millimeters. It may be read as 42.365783 square meters, or as 4,236.5783 square decimeters, or as 423,657.83 square centimeters, or as 42,365,783 square millimeters.

Now take 75.683256 cubic meters. It is composed of 75 cubic meters, 683 cubic decimeters, 256 cubic centimeters. It may be read as 75.683256 cubic meters, or as

75,683.256 cubic decimeters, or as 75,683,256 cubic centimeters.

The metric system is a system of related weights and measures. The meter is the basis from which all other units are derived. The unit of weight, the gram, is the weight of a cubic volume of water measuring 1 centimeter (.01 meter) on a side. The unit of capacity, the liter, is the volume of 1 kilogram (1,000 grams) of water and thus is represented by a cube measuring 1 decimeter (10 centimeters) on a side.

The liter and its derivatives are used for both dry and liquid measure.

In the following tables wherever the metric equivalents of U.S. standard measures are given, metric equivalents of other denominations may be found by simply moving the decimal point to the right or the left as may be necessary.

## Equivalent Values

### Linear Measure

| | | |
|---|---|---|
| 1 inch | = | 2.5400 centimeters |
| 1 foot | = | 0.3048 meter |
| 1 yard | = | 0.9144 meter |
| 1 mile | = | 1.6093 kilometers |
| 1 centimeter | = | 0.3937 inch |
| 1 decimeter | = | 3.9370 inches |
| 1 decimeter | = | 0.3281 foot |
| 1 meter | = | 39.3700 inches |
| 1 meter | = | 3.2808 feet |
| 1 meter | = | 1.0936 yards |
| 1 kilometer | = | 3,280.83 feet |
| 1 kilometer | = | 1,093.611 yards |
| 1 kilometer | = | 0.62137 miles |

### Square Measure

| | | |
|---|---|---|
| 1 square inch | = | 6.4516 square centimeters |
| 1 square foot | = | 0.0929 square meter |
| 1 square yard | = | 0.8361 square meter |
| 1 acre | = | 4,046.8730 square meters |
| 1 acre | = | 0.404687 hectare |
| 1 square mile | = | 258.9998 hectares |
| 1 square mile | = | 2.5900 kilometers |
| 1 square centimeter | = | 0.1550 square inch |
| 1 square decimeter | = | 15.5000 square inches |
| 1 square meter | = | 1,550.0000 square inches |
| 1 square meter | = | 10.7640 square feet |
| 1 square meter | = | 1.1960 square yards |
| 1 hectare | = | 2.4710 acres |
| 1 square kilometer | = | 247.1040 acres |
| 1 square kilometer | = | 0.3861 square mile |

The hectare is the unit of land measure.

### Capacity

| | | |
|---|---|---|
| 1 fluid dram | = | 3.6966 milliliters |
| 1 fluid ounce | = | 29.5730 milliliters |
| 1 liquid pint | = | 0.4732 liter |
| 1 liquid quart | = | 0.9463 liter |
| 1 gallon | = | 3.7853 liters |
| 1 milliliter | = | 0.2705 fluid dram |
| 1 milliliter | = | 0.0338 fluid ounce |
| 1 liter | = | 2.1134 liquid pints |
| 1 liter | = | 1.0567 liquid quarts |
| 1 liter | = | 0.2642 gallon |

| | | |
|---|---|---|
| 1 dry quart | = | 1.1012 liters |
| 1 dry peck | = | 0.8810 dekaliter |
| 1 bushel | = | 0.3523 hectoliter |
| 1 liter | = | 0.9081 dry quart |
| 1 dekaliter | = | 1.1351 pecks |
| 1 hectoliter | = | 2.8378 bushels |

The liter is used for both liquid and dry measure.

The milliliter is equivalent in volume to a cubic centimeter.

## Cubic Measure

| | |
|---|---|
| 1 cubic inch | = 16.3872 cubic centimeters |
| 1 cubic foot | = 28.3170 cubic decimeters |
| 1 cubic yard | = 0.7645 cubic meter |
| 1 cubic centimeter | = 0.0610 cubic inch |
| 1 cubic decimeter | = 0.0353 cubic foot |
| 1 cubic meter | = 1.3079 cubic yards |

## Weight

| | |
|---|---|
| 1 ounce troy | = 31.103 grams |
| 1 pound troy | = 0.3732 kilogram |
| 1 ounce avoirdupois | = 28.350 grams |
| 1 pound avoirdupois | = 0.4536 kilogram |
| 1 gram | = 0.0322 ounce troy |
| 1 gram | = 0.0353 ounce avoir-dupois |
| 1 kilogram | = 2.6792 pounds troy |
| 1 kilogram | = 2.2046 pounds avoir-dupois |

## Exercise Set 11.4

1. How many kilometers in 3,746.23 meters?
2. How many meters in 4.253 kilometers?
3. How many square kilometers in 85.46 square meters?
4. How many square millimeters in 47.386 square decimeters?
5. How many cubic centimeters in 3.56 cubic meters?
6. How many cubic meters in 374,658 cubic millimeters?
7. How many centiliters in 312.3765 liters?
8. How many hectoliters in 312.3765 liters?
9. How many centigrams in 7.46 kilograms?
10. How many kilograms in 3,426 grams?
11. Convert 385.25 hectares to acres.
12. How many bushels are there in 375 hectoliters?
13. How many miles in 153 kilometers?
14. How many gallons in 483 dekaliters?
15. Change 75.5 kilograms to pounds avoirdupois.
16. How many meters in 87 yards?
17. Change 157.35 acres to hectares.
18. Reduce 173 gallons to dekaliters.
19. How many hectoliters in 187 bushels?

## Chapter 11 Glossary

**Area** The number of square units in a figure.

**Base** The top or bottom of a figure.

**Cube** A rectangular solid in which all the faces are equal.

**Lateral Surface Area** The area of the sides of a solid figure, other than a rectangular one.

**Metric System** A system of related weights and measures based on decimal values.

**Plane Geometry** Geometry dealing with surfaces, or with figures having two dimensions.

**Rectangular Solid** A solid in which all the faces are rectangles.

**Solid Geometry** Geometry dealing with solids, or bodies having three dimensions.

**Volume** The number of cubic units in a solid.

## Chapter 11 Test

For each problem, five answers are given. Only one answer is correct. After you solve each problem, check the answer that agrees with your solution.

1. If a circle has a 3-inch radius, what is the area of the circle?

   **A)** 9.64 square inches
   **B)** 124.14 square inches
   **C)** 28.26 square inches
   **D)** 16.24 square inches
   **E)** 64.32 square inches

2. Using the figure below, find the volume of the rectangular solid.

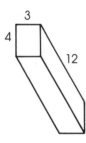

   **A)** 144 cubic inches
   **B)** 480 cubic inches
   **C)** 256 cubic inches
   **D)** 480 cubic inches
   **E)** 144 cubic inches

3. Find the area of a triangle with a base 6 inches long and with a height of 14 inches.

   **A)** 84 inches
   **B)** 42 square inches
   **C)** 66 square inches
   **D)** 84 square inches
   **E)** 28 square inches

4. If a trapezoid has one base measuring 12 feet, and another measuring 8 feet with a 4-foot height, what is the area?

   **A)** 54 square feet
   **B)** 36 square feet
   **C)** 24 square feet
   **D)** 20 square feet
   **E)** 40 square feet

5. What is the difference between 8 square feet and 8 feet square?

   **A)** 16 feet
   **B)** 16 square feet
   **C)** 56 square feet
   **D)** 64 square feet
   **E)** 48 square feet

6. A fence surrounding a ½ mile racetrack is 8 feet high. How many square yards of wood are contained in the fence?

   **A)** 5,180          **D)** 685⅔
   **B)** 586⅔          **E)** 945
   **C)** 2,346⅔

7. How many centimeters are in a meter?

   **A)** 50          **D)** 0.1
   **B)** 100          **E)** 10
   **C)** 1,000

8. How many meters in a kilometer?

   **A)** 10          **D)** 0.001
   **B)** 1,000          **E)** 100
   **C)** 0.01

9. How many grams are in a kilogram?

   **A)** 100          **D)** 0.1
   **B)** 10          **E)** 1,000
   **C)** 5

10. How many square meters are in a square kilometer?

   **A)** 100          **D)** 500
   **B)** 1,000,000          **E)** 100,000,000
   **C)** 10,000

11. How many liters are in a dekaliter?

   **A)** 100          **D)** 10
   **B)** 0.01          **E)** 1,000
   **C)** 0.001

12. If a box measures 12 inches by 5 inches by 16 inches, what is its volume?

    **A)** 690 cubic inches
    **B)** 960 cubic inches
    **C)** 840 cubic inches
    **D)** 540 cubic inches
    **E)** 780 cubic inches

13. If a circle has an area of 200.96 square inches, what is the diameter of the circle?

    **A)** 64 inches       **D)** 32 inches
    **B)** 24 inches       **E)** 8 inches
    **C)** 16 inches

14. What is the length of a rectangle with a 48-square-inch area, and a 4-inch width?

    **A)** 10 inches       **D)** 12 inches
    **B)** 16 inches       **E)** 24 inches
    **C)** 8 inches

15. What is the length of a side of a square with a 128-inch perimeter?

    **A)** 42 inches       **D)** 64 inches
    **B)** 24 inches       **E)** 16 inches
    **C)** 32 inches

16. What is the volume of a right cylinder with a 28.26-square-inch circular base, and a 17-inch height?

    **A)** 642.64 cubic inches
    **B)** 340.12 cubic inches
    **C)** 220.26 cubic inches
    **D)** 576.20 cubic inches
    **E)** 480.42 cubic inches

17. What is the volume of a pyramid with a 4-inch-square base and a 15-inch height?

    **A)** 80 cubic inches
    **B)** 110 cubic inches
    **C)** 45 cubic inches
    **D)** 60 cubic inches
    **E)** 90 cubic inches

18. What is the volume of a pyramid with a 36-square-inch triangular base and a 3-inch height?

    **A)** 78 cubic inches
    **B)** 108 cubic inches
    **C)** 90 cubic inches
    **D)** 120 cubic inches
    **E)** 96 cubic inches

19. What is the volume of a crate that measures 2 feet by 6 feet by 14 feet?

    **A)** 220 cubic feet
    **B)** 128 cubic inches
    **C)** 130 cubic inches
    **D)** 168 cubic inches
    **E)** 84 cubic inches

20. What is the height of a triangle with a 10-inch base and a 60-square-inch area?

    **A)** 12 inches       **D)** 6 inches
    **B)** 11 inches       **E)** 30 inches
    **C)** 5 inches

21. What is the cost of putting carpet down in a room 25 feet wide and 33½ feet long with carpet that costs $9.50 a square yard?

    **A)** $171.65       **D)** $1,500
    **B)** $950          **E)** $276.84
    **C)** $572

22. A pathway 50 feet long and 5 feet wide is paved with bricks measuring 8 inches by 4 inches. How many bricks are needed for the pathway?

    **A)** 2,250         **D)** 4,500
    **B)** 1,125         **E)** 2,933
    **C)** 3,357

# Graphs

## 12.1 Parts of a Graph

A **graph** is a diagram showing relationships between two or more factors. Most graphs have two scales, as the following population graph indicates:

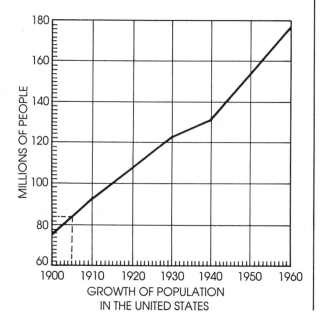

GROWTH OF POPULATION
IN THE UNITED STATES

In this graph, there is a vertical scale, and a horizontal scale. The vertical scale is sometimes called the vertical axis or ordinate, while the horizontal scale is called the horizontal axis or abscissa.

A **scale** is a convenient representation of one quantity or magnitude in terms of another. A scale expresses a ratio.

For instance, what covers miles in reality occupies only inches on a map. Therefore, a map has a scale based on the ratio between actual distances mapped and the distances on the map. This scale is usually shown on the map by a diagram, as illustrated below.

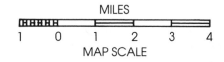

MILES

MAP SCALE

A thermometer has a scale based on the ratio between the height of a column of mercury and the temperature of air or of some other material, as measured in centigrade or Fahrenheit degrees.

A blueprint is drawn to scale. This means that the drawing on the blueprint represents a scale ratio. For instance, in the blueprint of the hull of a ship, if the scale is ¼ inch to 1 foot, then every line 1 inch in length on the blueprint represents 4 feet of the hull.

EXAMPLE 1: If the beam of a hull is to be 100 feet, how long a line would be needed on the blueprint just described in order to represent this beam?

SOLUTION: If the scale is ¼ inch to 1 foot, then 1 inch stands for 4 feet, and 1 foot stands for 48 feet.

We divide 100 by 4 to get the total number of inches necessary to draw the hull on the blueprint.

$$100 \div 4 = 25$$

We note that 25 inches converts to 2 feet, 1 inch.

### Exercise Set 12.1

1. If the scale on a blueprint is ½ inch to 1 foot, then how long is a wall 3 inches long on the blueprint?
2. If the scale on a blueprint is ⅕ inch to 2 inches, then how long is the blueprint drawing of a support beam 12 feet long?
3. If the scale on a blueprint is ¼ inch to 8 feet, then how high is a tree 4 inches high on the drawing?

4. If the scale on a blueprint is ⅔ inch to 3 feet, then how long is a 60-foot wall on the blueprint drawing?
5. If the scale on a blueprint is ¼ inch to 3 feet, then how long is a pipe that is 3 inches long on the blueprint?

## 12.2 Reading Graphs

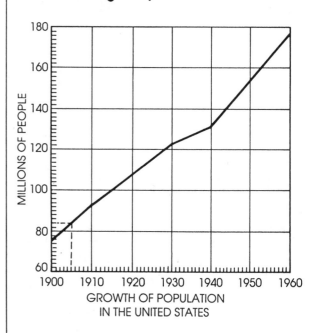

GROWTH OF POPULATION
IN THE UNITED STATES

This graph, which we saw first in Section 12.1, shows the population of the United States in millions from 1900 to 1960. If we could assume that the population growth was fairly steady for any ten-year period, then we could estimate the population for any year from the graph.

EXAMPLE 1: Find the population in the year 1905.

SOLUTION: We note that 1905 is halfway between 1900 and 1910. Move horizontally along the horizontal axis until we reach the tick mark that indicates the year 1905. From here, move straight up until we reach the graphed line. From here, move left until we reach the vertical axis. Make sure that the lines going up and across are straight.

The point where our line intersects the vertical axis is about ⅕ of the way between 80 and 100 millions. The population in 1905 would be about 84 million.

## Exercise Set 12.2

Use the graph from EXAMPLE 1 to answer the following questions.

1. How many millions of people were there in the United States in 1940?
2. In what year were there approximately 122 million people in the United States?
3. Approximately how many people were there in the United States in 1960?
4. In what year was the population of the United States approximately 152 million?
5. In what year was the population of the United States approximately 120 million?

## 12.3 Line Graphs

**Line graphs** are used to represent the factors in two different scales. Line graphs are particularly suitable for recording historical data involving a factor that constantly changes and fluctuates. This type of graph is preferred for presenting price records of stocks or commodities, or other changeable data.

This graph shows the variations in deaths due to car accidents between the years 1930 and 1990. The straight line segments are plotted from the actual number of accidents.

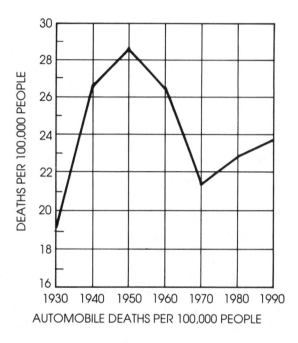

AUTOMOBILE DEATHS PER 100,000 PEOPLE

EXAMPLE 1: How many people per 100,000 died due to car accidents in 1970?

SOLUTION: Since we know the exact year, we begin by finding 1970 on the horizontal axis. We move straight up from the tick mark for 1970 until we meet the graphed line. When we meet the graph, we move horizontally to the left. When we reach the vertical axis, we find that we are between 21 and 22.

According to the data on this graph, about 21½ people per hundred thousand died in 1970. Since having a partial person killed does not make sense, we would round up to 22 people per hundred thousand killed.

To find out how many people died altogether in car accidents, we need to know the total population in 1970. If the total population in 1970 is 23,000,000, then we first divide 23,000,000 by 100,000.

$$23,000,000 \div 100,000 = 230$$

Then we multiply the quotient by our original result from EXAMPLE 1, $21\frac{1}{2}$.

$$230 \times 21\frac{1}{2} = 4,945$$

Thus, the total number of people killed in car accidents in 1970, according to our data, is 4,945 people.

Some line graphs represent quantities that change together. Such graphs may be made from formulas or equations.

EXAMPLE 2: Sam walks at a rate of 6 miles per hour. Show graphically the relation between the distance he walks, and the number of hours he walks.

SOLUTION: Use the formula $d = rt$. Since $r = 6$, we substitute to get $d = 6t$. We now make a table of hours and distances, using the formula.

| Hours | Distance |
|-------|----------|
| 1 | 6 |
| 2 | 12 |
| 3 | 18 |
| 4 | 24 |
| 5 | 30 |
| 6 | 36 |
| 7 | 42 |
| 8 | 48 |

Now we plot a graph by placing distances on one scale, and hours on another scale.

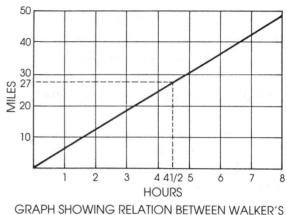

GRAPH SHOWING RELATION BETWEEN WALKER'S TIME AND DISTANCE

The range of numbers is decided according to the purpose of the graph. For hours the range selected is 8; for distance the range is 48. Next we choose our unit intervals for each scale. For the scale of hours, since the range is 8, we can choose one large or unit box on the graph paper as equal to 1 hour. For the scale of distance this is not practical, since we would need too much space. Therefore we choose a convenient interval. In this case let 1 small unit or square equal a distance of 10 miles.

Now, proceed to plot each point according to the pairs of items in the table. Thus the first point on the graph represents 1 hour and a distance of 6 miles, the second point represents 2 hours and a distance of 12 miles, etc. These points are called coordinates. When all the points are plotted draw the line, which in this case proves to be a straight line, showing that there must be a direct relationship between the distance covered and the time spent in walking.

Note: In plotting a graph which represents a formula or direct relation, it is not necessary to plot all the points. In a straight line graph three or four points are sufficient, two to determine the line, and one or two more as a check for accuracy.

From such a graph we may now read off directly the distance walked in 4½ hours.

First, 4½ would be midway between 4 and 5 on the hours' scale. Draw a line from this point to where it meets the line of the graph. From the point of the intersection draw a line out to the distance scale. This line meets the distance scale at 27, which is the answer. Check this by arithmetic in our formula. Substituting in $d = 6t$

$$d = 6 \times \frac{9}{2}$$
$$= \frac{54}{2}$$
$$= 27$$

## Exercise Set 12.3

Use the graph from EXAMPLE 1 to answer the following questions.

1. How many deaths per 100,000 people occurred in 1930?

2. How many deaths per 100,000 people occurred in 1950?

3. How many deaths per 100,000 people occurred in 1980?

4. How many deaths per 100,000 people occurred in 1940?

5. In what year did 29 deaths per 100,000 people occur?

6. In what year did 19 deaths per 100,000 people occur?

Use the graph from EXAMPLE 2 to answer the following questions.

7. If Sam walked for 3 hours, how far did he walk?

8. If Sam walked 40 miles, for how long did he walk?

9. If Sam walked for 1 hour, how far did he walk?

10. If Sam walked for 8 hours, how far did he walk?

11. If Sam walked 12 miles, how long did he walk?

## 12.4 Bar Graphs and Picture Graphs

**Bar graphs** use parallel bars to compare amounts of the same sort of measurement. For example, we can use bar graphs to compare the sales records of ten different sales representatives.

The bar graph below gives the total number of people examined each year at Lincoln Hospital from 1930 to 1980. The height of the black bar indicates the number of people examined. The base, or horizontal axis, indicates the specific year people were examined.

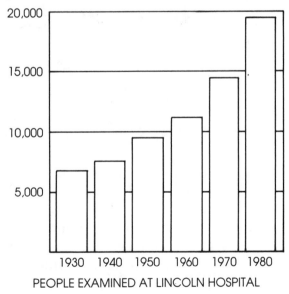

PEOPLE EXAMINED AT LINCOLN HOSPITAL

EXAMPLE 1: Approximately how many people were examined at Lincoln Hospital in 1960?

SOLUTION: Since we are given the year 1960, we begin by finding the bar for the year 1960. Then we move to the top of the bar for 1960, and move left to the vertical axis. We find the bar reaches about $\frac{1}{5}$ of the way up between 10,000 and 15,000. So we estimate that the number of people examined at Lincoln Hospital is about 11,000.

**Picture graphs** are used to illustrate statistical information, but the unit is not a distance measure on a page as in the line and bar graphs, but rather a symbol of the item being illustrated. The next figure illustrates such a graph in which several symbols are used.

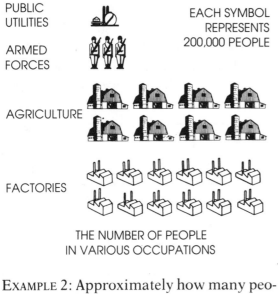

THE NUMBER OF PEOPLE
IN VARIOUS OCCUPATIONS

EXAMPLE 2: Approximately how many people have occupations in agriculture?

SOLUTION: First, we find the column for Agriculture. Then we count the number of symbols next to the word Agriculture. We count eight symbols, and since the graph notes that each symbol stands for 200,000 people, we multiply 8 by 200,000.

$$8 \times 200,000 = 1,600,000$$

Approximately 1,600,000 people have occupations in agriculture.

Picture graphs are less accurate sources of information because we must always round off numbers in order to draw a whole or $\frac{1}{2}$ symbol. Splitting up a symbol any further than into halves usually makes it too confusing to read the graph. When we must have more accurate information, we will use a line or a bar graph.

## Exercise Set 12.4

Use the graph from EXAMPLE 1 to answer the following questions.

1. In what year were approximately 14,000 people examined at Lincoln Hospital?
2. In what year were approximately 7,000 people examined at Lincoln Hospital?
3. In what year were approximately 19,000 people examined at Lincoln Hospital?
4. In which two years were approximately 10,000 people examined at Lincoln Hospital?
5. How many people were examined by Lincoln Hospital in 1940?

Use the graph from EXAMPLE 2 to answer the following questions.

6. How many people does each symbol represent?
7. How many people have occupations in the armed forces?
8. Which occupation employs 200,000 people?

**9.** How many people have occupations in factories?

**10.** Which occupation employs 600,000 people?

## 12.5 Circle Graphs

**Circle graphs** are sometimes called pie charts, as the divisions of the quantities look like slices of pie.

The circle graph is usually formed on a percentage basis. The whole circle represents 100%, while fractional percentages are indicated by proportionate segments of the circle.

The circle graph below shows a breakdown of expenses for Dan's Candy Shop. Each expense is presented as a percentage of the total expense.

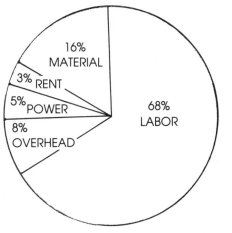

EXPENSES FOR DAN'S CANDY SHOP

EXAMPLE 1: If Dan's total expenses are $5,000, how much does he pay for rent?

SOLUTION: We must first find the percentage of the total expenses owed for rent. On the circle graph, 3% of the expenses are for rent. To find the amount Dan paid for rent, we multiply 3% by $5,000.

$$3\% \times \$5,000 = \$150$$

Dan pays $150 for rent.

## Exercise Set 12.5

Use the graph from EXAMPLE 1 to answer the following questions.

**1.** If expenses total $6,000, how much does Dan spend on power?

**2.** If total expenses equal $4,000, which expense costs approximately $133?

**3.** If total expenses equal $7,000, which expense costs approximately $583?

**4.** Which expense costs approximately $653, if total expenses equal $4,000?

**5.** Which expense costs approximately $2,666, if total expenses equal $4,000?

Make a circle graph to solve the following problem.

**6.** Julia's day is divided up as follows: she works 7 hours, eats and dresses for $3\frac{1}{2}$ hours, studies for 3 hours, travels $1\frac{1}{2}$ hours, plays tennis 1 hour, and sleeps 8 hours. Find the percentages for each in a 24-hour day, and draw a circle graph to illustrate.

## Chapter 12 Glossary

**Bar Graph** A graph that uses parallel bars to compare amounts of the same sort of measurement.

**Circle Graph** A graph that shows percentages of a whole by indicating proportionate segments of the circle.

**Graph**   A diagram showing relationships between two or more factors.

**Line Graph**   A graph that connects the points that represent the factors in two different scales with a line.

**Picture Graph**   A graph where symbols are used to illustrate statistical information.

**Scale**   A representation of one quantity or magnitude in terms of another quantity.

## Chapter 12 Test

For each problem, five answers are given. Only one answer is correct. After you solve each problem, check the answer that agrees with your solution.

Use this graph to answer questions 1 to 5.

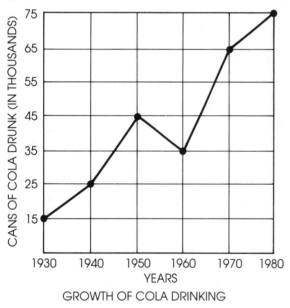

GROWTH OF COLA DRINKING

1. In which year were 75,000 cans of cola drunk?

   **A)** 1980
   **B)** 1975
   **C)** 1930

   **D)** 1945
   **E)** 1960

2. How many cans of cola were drunk in 1940?

   **A)** 15,000
   **B)** 65,000
   **C)** 50,000
   **D)** 12,000
   **E)** 25,000

3. How many cans of cola were drunk in 1970?

   **A)** 12,000
   **B)** 65,000
   **C)** 55,000
   **D)** 25,000
   **E)** 35,000

4. In what year were 45,000 cans of cola drunk?

   **A)** 1931
   **B)** 1960
   **C)** 1950
   **D)** 1980
   **E)** 1970

5. In what year were 15,000 cans of cola drunk?

   **A)** 1980
   **B)** 1930
   **C)** 1970
   **D)** 1960
   **E)** 1950

Use the following graph to answer questions 6 to 10.

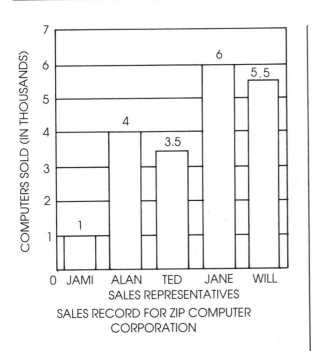

SALES RECORD FOR ZIP COMPUTER CORPORATION

**6.** How many computers did Ted sell?

**A)** 2,000
**B)** 3,500
**C)** 5,000
**D)** 7,500
**E)** 8,000

**7.** Who sold 6,000 computers?

**A)** Jami
**B)** Alan
**C)** Ted
**D)** Jane
**E)** Will

**8.** Who sold 4,000 computers?

**A)** Jami
**B)** Alan
**C)** Ted
**D)** Jane
**E)** Will

**9.** How many computers did Jami sell?

**A)** 2,000

**B)** 7,000
**C)** 5,000
**D)** 1,000
**E)** 3,000

**10.** Who sold 5,500 computers?

**A)** Jami
**B)** Alan
**C)** Ted
**D)** Jane
**E)** Will

Use the following graph to answer questions 11 to 15.

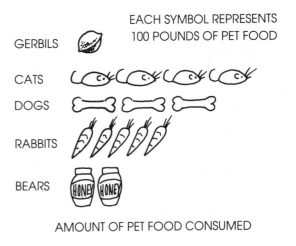

AMOUNT OF PET FOOD CONSUMED

**11.** How many pounds of pet food do cats consume?

**A)** 100
**B)** 300
**C)** 500
**D)** 400
**E)** 700

**12.** Which type of pet consumes 200 pounds of food?

**A)** Cats
**B)** Dogs
**C)** Rabbits

**D)** Bears

**E)** Gerbils

13. How many pounds of food do rabbits consume?

**A)** 200

**B)** 500

**C)** 300

**D)** 700

**E)** 100

14. Which type of pet consumes 500 pounds of food?

**A)** Cats

**B)** Dogs

**C)** Rabbits

**D)** Bears

**E)** Gerbils

15. Which type of pet consumes 100 pounds of food?

**A)** Cats

**B)** Dogs

**C)** Rabbits

**D)** Bears

**E)** Gerbils

Use the following graph to answer questions 16 to 20.

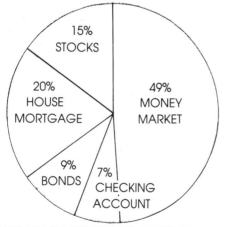

BARBARA'S DISTRIBUTION OF HER SAVINGS

16. What percentage of Barbara's money is invested in stocks?

**A)** 49%

**B)** 15%

**C)** 20%

**D)** 9%

**E)** 7%

17. If Barbara save $3,000, how much money goes to bonds?

**A)** $350

**B)** $2,000

**C)** $150

**D)** $420

**E)** $270

18. What percentage of Barbara's money goes to her checking account?

**A)** 49%

**B)** 15%

**C)** 20%

**D)** 9%

**E)** 7%

19. If Barbara saves $4,000, how much goes to stocks?

**A)** $300

**B)** $700

**C)** $600

**D)** $200

**E)** $100

20. What percentage of Barbara's money goes to her house mortgage?

**A)** 49%

**B)** 15%

**C)** 20%

**D)** 9%

**E)** 7%

# Probability

## 13.1 Permutations

When the coach assigns nine of the twenty members of a baseball squad to bat in a particular batting order, mathematically the coach may be said to be making a permutation of twenty things nine at a time.

**Permutations** are arrangements of things in a particular order.

The symbolic representation for the total number of possible arrangements of $n$ things taken $t$ at a time is:

$P(n,t)$ or $P_t^n$

The notation used to designate all the possible permutations of twenty things taken nine at a time is:

$$P(20,9) \quad \text{or} \quad P_9^{20}$$

As an illustration, suppose that we have three paintings, labeled $A$, $B$, and $C$. We want to arrange all three of the paintings on a wall. We wish to find the number of permutations of three paintings all hung on the wall.

$$P(3,3) = 6 \quad ABC, BAC, CAB,$$
$$ACB, BCA, CBA$$

Notice that each permutation contains all three paintings. The only difference between the permutations is the order in which the paintings are arranged.

If we wish to find the number of permutations of the three paintings when only two of the paintings will be hung on the wall, we write:

$$P(3,2) = 6 \qquad AB, BA, CA, AC,$$
$$BC, CB$$

Notice that each permutation contains only two of the three paintings. The order of the paintings is different in each of the permutations.

Note that each of the permutations above begins with one of the three possibilities, *A*, *B*, or *C*. But once this first permutational position has been filled, only two possibilities remain to be assigned to the second permutational position. Either *B* or *C* can follow *A; A* or *C* can follow *B*; or *A* or *B* can follow *C*. Once the first two permutational positions have been filled, the one remaining possibility can be assigned to the third permutational position.

We see that $P(3,2)$ must be the product of the three possible entries for the first permutational position, times $(3 − 1) =$ two possible entries for the second permutational position; or $3(2) = 6$. This product agrees with our result above.

Thus we also see that $P(3,3)$ must be the product of the three possible entries for the first permutational position, times $(3 − 1)$ = two possible entries for the second permutational position, times $3 − 2 =$ one possible entry for the third permutational position; or $(3)(2)(1) = 6$, which also agrees with our result.

There are an equal number of ways of listing these two different permutations because any number multiplied by 1 remains the same number. $P(3,3)$ and $P(3,2)$ are equal.

The product written out in the form $3(2)(1)$ is called three **factorial.** The symbol for it is 3!, meaning 3 times 2 times 1.

A factorial is a product symbolized by $n!$, which means

$$n! = n(n − 1)(n − 2) \cdots (3)(2)(1)$$

The value of 0! is defined to be 1. That is because it is assumed that there is one way to arrange zero objects. The value of 1! is also defined to be 1. It is assumed also that there is one way to arrange one object.

The total number of permutations of any *n* things taken *t* at a time is given in the formula:

$$P(n,t) = \frac{n!}{(n − t)!}$$

In the special case where $t = n$, this becomes:

$$P(n,n) = n!$$

since the denominator of the $P(n,t)$ formula is then eliminated.

Permutations formulas are applicable to the solution of many different types of problems involving ordered arrangements.

EXAMPLE 1: In how many ways can five of seven different wires from one piece of electrical equipment be attached to five different posts on another piece of equipment?

SOLUTION: Since the problem involves order, the problem involves permutations, with $n = 7$, and $t = 5$. Use the permutations formula to solve the problem.

$$P(7,5) = \frac{7!}{(7-5)!}$$

$$= \frac{7!}{2!}$$

$$= \frac{7(6)(5)(4)(3)(2)(1)}{2(1)}$$

$$= 2,520$$

Again, we can see from the arithmetic performed on this solution that the factor of 1 never affects the result. We shall leave it implied (not written) from now on in our numerical computations.

EXAMPLE 2: In how many different arrangements may five people be seated along the head of a banquet table?

SOLUTION: Here is a permutation problem, where $n = t = 5$.

$$P(5,5) = 5!$$

$$= 5(4)(3)(2) \quad \text{remember that} \atop \text{1 is implied}$$

$$= 120$$

When special conditions are attached to permutation problems, the same formulas may often still be used, but they must be adapted to the different requirements of the problems.

EXAMPLE 3: How many different arrangements are there possible in the preceding example if one couple among the five persons must always be seated next to one another?

SOLUTION: If we begin by thinking of the couple as a single unit, then we may begin by considering the initial part of the problem as one of permuting this unit with the three other persons. Then $n = t = 4$, and

$$P(4,4) = 4! = 4 \cdot 3 \cdot 2 = 24$$

But in each such permutation, the two members of the couple may sit next to each other in either of the two possible orders, $AB$ or $BA$. Therefore, the total number of possible permutations is really twice $P(4,4)$, or

$$2P(4,4) = 2(24) = 48$$

EXAMPLE 4: How many different arrangements are possible if the couple must always be seated apart from one another?

SOLUTION: The entire group can be seated with no special conditions in $P(5,5)$ different ways, and with the special pair always together in $2P(4,4)$ different ways. Therefore, the number of ways in which the five people can be seated with the couple separated must be the difference between these two figures, or

$$P(5,5) - 2P(4,4) = 120 - 48 = 72$$

EXAMPLE 5: How many ways may the same five people be seated around a round table?

SOLUTION: In a circular arrangement, no one person is actually first, last, or centered. But we may arbitrarily think of any one person's position as fixed for purposes of reference and then permute the positions of the other four with respect to each other, clockwise or counterclockwise from this point of reference. Thus $n = 5 - 1 = 4$, and then as above

$$P(4,4) = 4! = 24$$

Arrangements of things in a closed chain or ring as in the last example are called circular permutations. Generalizing the method of reasoning just illustrated, we define the number of permutations of $n$ distinct objects arranged in a circle as $P_c = (n - 1)!$.

Therefore, the preceding solution could have been arrived at directly, by substitution into this formula, as

$$(5 - 1)! = 4! = 24$$

EXAMPLE 6: Suppose we have five people waiting in line to see a movie. If three of the five people waiting are men, and two are women, in how many ways can they wait in line if they are distinguished only by gender, and not as individuals?

SOLUTION: Let our answer be $P_a$ the total number of permutations possible for three indistinguishable men, and two indistinguishable women, taken all five at a time. Then, in order to express $P(5,5)$ in terms of $P_a$ we would have to multiply $P_a$ by $P(3,3) = 3!$ to obtain the number of permutations

in which the three men are differentiated as $M_1, M_2, M_3$, and by $P(2,2) = 2!$ to obtain the number of cases in which the two women are also differentiated as $W_1, W_2$. But we already know that $P(5,5) = 5!$. By substitution we get

$$3!2!P_a = P(5,5) = 5!$$

And therefore

$$P_a = \frac{5!}{3!2!} = \frac{5 \cdot \overset{2}{\cancel{4}} \cdot \overset{1}{\cancel{3}} \cdot \overset{1}{\cancel{2}}}{\underset{1}{\cancel{3}} \cdot \underset{1}{\cancel{2}} \cdot \underset{1}{\cancel{2}}} = 10$$

These ten possible ways, six beginning with a man and four with a woman, are illustrated as follows:

$$6 \begin{cases} MMMWW \\ MMWMW \\ MMWWM \\ MWMMW \\ MWMWM \\ MWWMM \end{cases} \quad \left. \begin{matrix} WMMMW \\ WMMWM \\ WMWMM \\ WWMMM \end{matrix} \right\} 4$$

Note that the preceding problem requires us to treat its "men" as $M$'s "all alike" and its "women" as $W$'s "all alike." This is doubtless sociologically superstitious and matrimonially foolish. But the accompanying mathematical procedure is nevertheless sound. Generalizing the solution's method of reasoning, we derive for $P_a$, the total number of **indistinguishable permutations** possible for $n$ things taken $n$ at a time when $n_1$ are all alike, and $n_2$ are all alike, etc., the formula,

$$P_a = \frac{n!}{n_1!n_2! \cdots}$$

EXAMPLE 7: How many indistinguishable nine-letter words can be formed by permutations of the letters in the word "TENNESSEE"?

SOLUTION: Here $n = 9$, $n_1 = 4$ for the $E$ letters, and $n_2 = n_3 = 2$ for the $N$ and $S$ letters. Hence,

$$P_a = \frac{9!}{4!2!2!}$$

$$= \frac{\overset{2}{\cancel{9} \cdot \cancel{8}} \cdot 7 \cdot 6 \cdot 5 \cdot \overset{1}{\cancel{4}} \cdot \overset{1}{\cancel{3}} \cdot \overset{1}{\cancel{2}}}{\underset{1}{\cancel{4}} \cdot \underset{1}{\cancel{3}} \cdot \underset{1}{\cancel{2}} \cdot \underset{1}{\cancel{2}} \cdot \underset{1}{\cancel{2}}}$$

$$= 3{,}780$$

EXAMPLE 8: In how many different ways can a coach assign speakers to the first and second positions of a two-man debating team from a squad of twelve candidates?

SOLUTION: Here $N = 12$, $t = 2$, and

$$P(12,2) = \frac{12!}{(12 - 2)!} = \frac{12!}{10!}$$

$$= \frac{12 \cdot 11 \cdot \cancel{10} \cdot \cancel{9} \cdot \cancel{8} \cdot \cancel{7} \cdot \cancel{6} \cdot \cancel{5} \cdot \cancel{4} \cdot \cancel{3} \cdot \cancel{2}}{\cancel{10} \cdot \cancel{9} \cdot \cancel{8} \cdot \cancel{7} \cdot \cancel{6} \cdot \cancel{5} \cdot \cancel{4} \cdot \cancel{3} \cdot \cancel{2}}$$

$$= 132$$

Or, with the second line written more briefly,

$$= 12 \cdot 11 \cdot \frac{10!}{10!} = 132$$

since $10!/10! = 1$

From this example we see that it is always correct, and often arithmetically more convenient, to treat

$$\frac{n!}{(n - t)!} = n(n - 1)(n - 2) \cdots (n - t + 1)$$

in the basic formula for $P(n, t)$. From now on, therefore, we shall write $12!/(12 - 10)!$ or $12!/10!$ directly as $12 \cdot 11$, etc., whenever we do not wish to use the other factors of the numerator for cancellations with other denominator factors in the same term.

## Exercise Set 13.1

1. Compute $P(4,3)$.
2. How many different signals can we make by hoisting five different flags in a line to a ship's masthead?
3. How many different signals can we make by hoisting five of twelve different flags in a line to the ship's masthead?
4. In how many ways can eight people be seated at a round table if one particular couple cannot be separated?
5. How many five-letter "words" can there be formed, such as "crate" or "write," where no letter is repeated, the third and fifth letters are from the five vowels, *a, e, i, o, u,* and the other three are from the other twenty-one letters of the alphabet?
6. In how many different indistinguishable sequences can three pennies, six nickels, and four dimes be arranged?
7. Find the number of four-digit numbers possible using the digits 0, 1, 2, 3, if each digit can only be used once.
8. Compute $P(5,4)$.
9. Compute $P(8,2)$.
10. Compute $P(7,1)$.

## 13.2 Combinations

When the baseball coach picks nine players out of a squad of twenty players to form a team, mathematically the coach can be said to be making a combination of twenty "things" taken nine at a time. **Combinations** are groupings of things without regard to their order.

---

The common symbols for the total number of all possible combinations of $n$ things taken $t$ at a time are:

$C(n,t)$   or   $C_t^n$

---

Here is the notation used to designate all the possible combinations of twenty things taken nine at a time:

$$C(20,9) \quad \text{or} \quad C_9^{20}$$

For any given pair of numerical values of $n$ and $t$ in these formulas, there are, in general, many more possible permutations than there are possible combinations.

---

The formula for combinations is:

$$C(n,t) = \frac{n!}{t(n-t)!}$$

---

The formula for combinations can be applied to the solution of many different types of problems in which selections are to be made without regard to order.

EXAMPLE 1: Find out how many ways there are to combine three out of three paintings on a wall.

SOLUTION: There is only one possible grouping of all three of the paintings that we can select at one time without regard to order.

$$C(3, 3) = \frac{3!}{3!(3-3)!}$$

$$= \frac{3!}{3!(0!)}$$

$$= \frac{(3)(2)(1)}{(3)(2)(1)(1)}$$

$$= 1$$

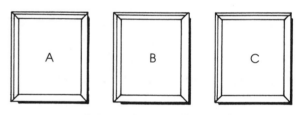

In a combination, the elements matter, not the order they are listed in. In this example, any other listing of the three paintings would be the same combination, no matter what order the paintings were listed in.

EXAMPLE 2: How many ways are there to combine two of three paintings on a wall?

SOLUTION: Here we are selecting groupings from the same three objects two at a time.

$$C(3, 2) = \frac{3!}{2!(3-2)!}$$

$$= \frac{3!}{2!(1!)}$$

$$= \frac{(3)(2)(1)}{(2)(1)(1)}$$

$$= 3$$

There are three ways to combine two of the three paintings at a time on a wall.

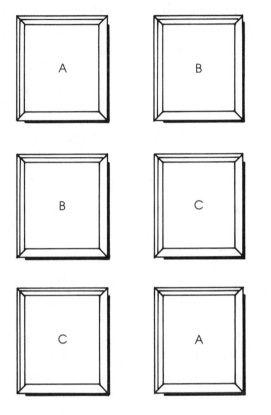

EXAMPLE 3: By forming arrays of all the possibilities, as above, find $C(5, 2)$ and $P(5, 2)$.

SOLUTION: Letting $A$, $B$, $C$, $D$, $E$ represent the five things, we can systematically combine them two at a time in any of these ways:

$C(5, 2)$:

$$\overparen{A \& B}$$

$$\overparen{A \& C}, \ \overparen{B \& C}$$

$$\overparen{A \& D}, \ \overparen{B \& D}, \ \overparen{C \& D}$$

$$\overparen{A \& E}, \ \overparen{B \& E}, \ \overparen{C \& E}, \ \overparen{D \& E}$$

And we can systematically permute these combinations in these ways:

$P(5, 2)$:

| AB | BA | CA | DA | EA |
|----|----|----|----|----|
| AC | BC | CB | DB | EB |
| AD | BD | CD | DC | EC |
| AE | BE | CE | DE | ED |

By actual count of these arrays, as in the preceding illustrations, we now find that

$$C(5, 2) = 10, P(5, 2) = 20$$

In other words there are ten possible combinations of five things taken two at a time; and there are twenty possible permutations of five things taken two at a time.

Unless $n$ and $t$ have very small values, it is tedious to find $C(n, t)$ and $P(n, t)$ by the method of the above illustrations in which all the separate possibilities are written out in an array and then counted. Therefore, in actual practice, short-method formulas are used.

From the very beginning of this chapter we have seen that permutations are simply different arrangements of the things grouped in corresponding combinations. More particularly, each set of permutations of any given $t$ things in $P(n, t)$ consists simply of different arrangements of the same $t$ things in a corresponding combination of $C(n, t)$.

But now we know from our basic permutations formula that each combination of $t$ things in $C(n, t)$ can have $P(t, t) = t!$ possible permutations:

$$t!C(n, t) = P(n, t)$$

Dividing by $t! = t!$, we then obtain the second important basic relationship

$$C(n, t) = \frac{P(n, t)}{t!}$$

and by substitution for $P(n, t)$ from our basic permutations formula, this gives us for the number of all possible combinations of $n$ things taken $t$ at a time, the basic formula:

$$C(n, t) = \frac{n!}{t!(n - t)!}$$

In the special case where $t = n$, the denominator factor, $(t - t)!$ or $(0)!$ is again eliminated, and the above formula takes the anticipated form

$$C(n, n) = n!/n! = 1$$

Combinations formulas can be applied to the solution of many different types of problems in which selections are to be made without regard to the possible permutations of arrangement.

EXAMPLE 4: In how many different ways can a poker hand of five cards be dealt from a fifty-two-card deck?

SOLUTION: Since the grouping of five cards is important, and the order of the cards is not important, the problem is a combinations problem. Here let $n = 52$, and let $t = 5$, and use the combinations formula to solve the problem.

$$C(52, 5) = \frac{52!}{5!(52 - 5)!}$$

$$= \frac{52!}{5!47!}$$

$$= \frac{(52)(51)(50)(49)(48)47!}{5!47!}$$

Notice that we can simplify the combinations formula here without having to multiply out all of the terms of the product. Note that $\frac{47!}{47!}$ is equal to 1. Rewrite the product now as:

$$= \frac{(52)(51)(50)(49)(48)}{5!}$$

$$= \frac{(52)(51)(50)(49)(48)}{(5)(4)(3)(2)(1)}$$

$$= 2,598,960$$

There are 2,598,960 different possible five-card poker hands that can be dealt from a fifty-two-card deck.

EXAMPLE 5: (a) In how many different ways may a coach assign players, from a squad of twelve, to the nine positions on a baseball line-up? (b) In how many different ways can he select a team of nine from the same squad without regard to the positions they are to play?

SOLUTION: Here $n = 12$ and $t = 9$. Since question (a) involves positions, we must again use the previously derived permutations formula:

$$\textbf{a)} \ \ P(12, 9) = \frac{12!}{(12 - 9)!} = \frac{12!}{3!}$$

$$= 12 \cdot 11 \cdot 10 \cdot 9 \cdot 8$$

$$\cdot 7 \cdot 6 \cdot 5 \cdot 4$$

$$= 79,833,600$$

But since question (b) concerns selection without regard to position played, we must now use our new combinations formula:

$$\textbf{b)} \ \ C(12, 9) = \frac{12!}{9!(12 - 9)!} = \frac{12!}{9!3!}$$

Then, dividing numerator and denominator by 9! rather than by 3!, this gives us most simply:

$$\frac{\overset{4}{\cancel{12}} \cdot 11 \cdot \overset{5}{\cancel{10}}}{\underset{1}{\cancel{3}} \cdot \underset{1}{\cancel{2}}} = 220$$

Note in this numerical instance how very large a number of possible permutations may be as compared with the corresponding number of possible combinations. The latter may also be surprisingly large, however; and then it is important to shorten the computation by the arithmetic device of the above solution.

Thus far we have meant by "combinations" only "subgroup combinations"—those of $t$ things taken from the same group of $n$ things. But many related problems concern **intergroup combinations.** These may be defined as selections of $t_1$ things at a time from a first group of $n_1$ things, along with $t_2$, things at a time from a second group of $n_2$ things, along with $t_3$ things at a time from a third group of $n_3$ things, etc.

Suppose, for instance, that we wish to combine one of two photographs, $R$ and $S$, with two of the three paintings, $A$, $B$, and $C$, mentioned at the beginning of this chapter. Here $n_1 = 2$; and the total number of possible intergroup combinations, $N = 6$, is illustrated by the array:

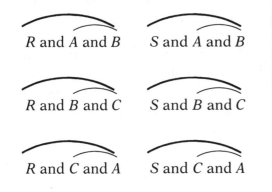

R and A and B     S and A and B

R and B and C     S and B and C

R and C and A     S and C and A

Evidently, here, we may make our first choice (of one of the two photographs, $R$ and $S$) in either of these two ways:

$$C(2, 1) = \frac{2!}{1!(2-1)!}$$
$$= \frac{2!}{1!1!} = \frac{2}{1 \cdot 1} = 2$$

Also, we may make our second choice (of two of the three paintings, $A$, $B$, and $C$) in any of these three ways:

$$C(3, 2) = \frac{3!}{2!(3-2)!}$$
$$= \frac{3!}{2!1!} = \frac{3 \cdot \overset{1}{\cancel{2}}}{\underset{1}{\cancel{2}}} = 3$$

Hence, the total number, $N$, of possible ways in which we may combine both of these choices (to make up a selection of one photograph and two paintings) is

$$N = C(2, 1)C(3, 2) = 2(3) = 6$$

Applying the same method of reasoning to the general case defined above, therefore, we derive the **basic intergroup combinations** formula.

> $N = C(n_1, t_1)C(n_2, t_2) \cdots C(n_k, t_k),$
>
> for the total number of joint selections possible from $k$ different sets of $n_1$, $n_2$ $\cdots n_k$, things taken $t_1$, $t_2 \cdots t_k$ at a time, respectively. In the special case where $t_1 = t_2 = \cdots t_k = 1$, this becomes
>
> $N = n_1 n_2 \cdots n_k$

$C(n, 1) = n$ for all possible values of $n$.

The $k$ different kinds of things from which choices are to be combined in applications of these formulas may be objects, events, placements, or whatever. They may also occur or be selected in the actual time sequence, $1, 2, 3 \cdots k$, or simultaneously. In the latter case, however, it is convenient to assign some one arbitrarily selected order to them. This is to avoid the sort of duplicated count we would get if we were (mistakenly) to regard the combination of photograph $R$ with paintings $A$ and $B$, in the above instance, as different from the combination of paintings $A$ and $B$ with photograph $R$ (the latter being a different permutation, but not a different combination).

EXAMPLE 6: In planning a plane trip from Boston to San Diego with stops at Chicago and St. Louis, a travel agent finds he can book a seat for his customer on five different flights to Chicago, on four different flights from there to St. Louis, and on seven different flights from there on. How many different complete itineraries can he book?

SOLUTION: Here $k = 3$ (different stages of the trip), and it is common-sense routine to let $n_1 = 5$ (possible choices for the first stage), $n_2 = 4$ (possible choices for the second stage), and $n_k = n_3 = 7$ (possible choices for the third stage), all in their actual time sequence. Then the total number of possible itineraries is

$$N = n_1 n_2 n_3 = 5(4)(7) = 140$$

EXAMPLE 7: An international scientific conference is attended by five representatives from North America, four from South America, seven from Europe, six from Asia, and three from Africa. How many different committees can be chosen from this assemblage with three representatives from each area?

SOLUTION: Here there is no natural time sequence of choice, but we may arbitrarily let $n_1 = 5$ (possibilities of choice from North America), $n_2 = 4$ (from South America), etc. Then $t_1 = t_2 = \cdots t_5 = 3$, and

$$N = C(5, 3)C(4, 3)C(7, 3)C(6, 3)C(3, 3)$$

$$= \frac{5 \cdot 4 \cdot 3}{3 \cdot 2} \cdot \frac{4 \cdot 3 \cdot 2}{3 \cdot 2} \cdot \frac{7 \cdot 6 \cdot 5}{3 \cdot 2} \cdot \frac{6 \cdot 3 \cdot 4}{3 \cdot 2} \cdot \frac{3 \cdot 2}{3 \cdot 2}$$

$$= (10)(4)(35)(20)(1) = 28,000$$

EXAMPLE 8: Designate the "head" and "tail" faces of three coins in any arbitrary order as $H_1$ and $T_1$, $H_2$ and $T_2$, $H_3$ and $T_3$, respectively. If we count $H_1 T_2 T_3$ as different from $T_1 H_2 T_3$ or $T_1 T_2 H_3$, etc., in how many different patterns, of their head or tail faces up, can the three coins land when tossed?

SOLUTION: Here the three coins and their corresponding faces may be identical. However, when we consider them in the proposed arbitrary order, $n_1 = 2$ (possibilities of $H_1$ or $T_1$ for the first coin), $n_2 = 2$ (possibilities of $H_2$ or $T_2$ for the second coin), etc.; then $k = 3$, and the total number of different possible ways in which the three coins can land is

$$N = n_1 n_2 n_3 = 2(2)(2) = 2^3 = 8$$

EXAMPLE 9: Each die of a pair has its faces numbered from one to six. In how many different patterns of their numbered faces up—"different" being defined as in the preceding problem—can the pair come to rest when rolled?

SOLUTION: By the method of the preceding solution, $k = 2$, and $n_1 = n_2 = 6$. Therefore:

$$N = n_1 n_2 = 6^2 = 36$$

From the last two examples we see that, in the further special case where $n_1 = n_2 = \cdots n_k = n$, the preceding formula becomes:

$$\boldsymbol{N} = n^k$$

In EXAMPLE 8, for instance, $k = 3$, $n = 2$, and

$$N = n^k = 2^3 = 8 \text{ (as before)}$$

In EXAMPLE 9, also, $k = 2$, $n = 6$, and

$$N = n^k = 6^2 = 36 \text{ (as before)}$$

But a more general result now follows. . . .

In a fairly common type of combinations problem we are required to find $M$, the sum of all the possible combinations of $k$ things taken 1, 2, 3 . . . and any other number up to, and including, $k$ at a time. This can always be done "the long way" by substitution in the defining formula

$$\boldsymbol{M} = C_1^k + C_2^k + \cdots + C_k^k$$

Note, however, that we can think of the $k$ things in this formula as the same $k$ as in the preceding formula for $N = n^k$. Also,

we can think of $n = 2$ in the latter as the two possibilities of each of the same $k$ things being either (a) included in possible combinations, or (b) excluded from such combinations. Then

$$N = M + 1, \quad or \quad M = N - 1$$

with the 1 in these equations being the number of the one possibility in which all $k$ things are excluded from any possible combination. Hence, by substitution, we obtain the formula

$$M = C_1^k + C_2^k + \cdots + C_k^k = 2^k - 1$$

(*Note:* Most textbooks derive this formula by a longer method which applies a special case of quite a different principle called the binomial theorem. But this method is recommended as shorter, more direct, and less dependent upon other principles or the special device of ingenious substitution.)

EXAMPLE 10: In how many different combinations can we replace one or more electronic tubes in a set of eight?

SOLUTION: Since these replacements may be made for any one, or two, or three, or more tubes up to and including all eight, our $M$ formula applies with $n = 8$. Therefore:

$$M = 2^8 - 1 = 256 - 1 = 255$$

The minus one in this computation adjusts the total for the rejected possibility that no tube be replaced.

As when we apply permutations formulas, we may sometimes have to adapt the above combination formulas to the special conditions of particular problems.

Care must be taken, however, not to insinuate duplications of arrangement in this procedure, for then we shall obtain numerical results which are too large by some factor of a corresponding permutations formula. The very real danger of this becomes immediately evident when we recognize that the reasoning by which we originally arrived at our basic permutations formula was itself but an application of our basic intergroup combinations formula to the systematic filling of successive positions in a permutational arrangement.

EXAMPLE 11: Of all the possible poker hands computed in the solution of EXAMPLE 4, how many contain "four of a kind"—defined as four cards of one rank (all sevens, or all jacks, etc.)?

SOLUTION: Since there are only four cards in each of the thirteen ranks (ace, king, queen, jack, ten, nine, eight, seven, six, five, four, three, two) of a bridge or poker deck, four cards of the same rank can be dealt from such a deck in only

$$13C_4^4 = 13 \cdot 1 = 13 \text{ different ways}$$

But each such combination can be further combined with any one of the remaining $12C_1^4 = 12 \cdot 4 = 48$ cards of 12 other ranks in the deck. Hence, the number of five-card hands with "four of a kind" is

$$13C_4^4 12C_1^4 = 13 \cdot 12 \cdot 4 = 624$$

EXAMPLE 12: In how many different ways can a poker hand be dealt containing only "one pair"—defined as two cards of one rank and three other cards of ranks different from each other and from that of the pair?

SOLUTION: Obviously a pair can be formed in each rank in $C_2^4$ ways, and hence in all thirteen ranks of the deck in

$$13C_2^4 = 13 \cdot \frac{4 \cdot 3}{2} = 78 \text{ ways}$$

---

**Warning**

There is a danger we may now make the mistake of reasoning (erroneously) that each such combination can be further combined with one card from each of three remaining ranks in

$$12C_1^4 11C_1^4 10C_1^4 = 12 \cdot 4 \cdot 11 \cdot 4 \cdot 10 \cdot 4$$
$$= 84,480 \text{ different ways}$$

Multiplied by 78, this would give us a product of 6,589,440 ways, which is impossible since it is more than twice as great as the total of all possible poker hands, including those with no pairs or with other subcombinations than pairs.

---

The error of the preceding reasoning is that it permutes, rather than combines, selections from three of the remaining suits. Hence, the product at which it arrives is too great by a factor of $3! = 3 \cdot 2 = 6$.

To correct the error, we could divide the obtained product by a compensating factor of $3! = 6$. Or, to avoid the error altogether, we can reason that three ranks different from the pair's rank can be combined in

$$C_3^{12} = \frac{12 \cdot 11 \cdot 10}{3 \cdot 2} = 220 \text{ ways}$$

Then from each of the latter we can make an intergroup combination of one of $n = 4$ cards from each of $k = 3$ ranks in

$$n^k = 4^3 = 64 \text{ ways}$$

Hence a five-card hand with only "one pair" can be dealt in

$$13C_2^4 64C_3^{12} = 78 \cdot 64 \cdot 220$$
$$= 1{,}098{,}240 \text{ ways}$$

The most serious problem to avoid in applying these formulas is that of erroneously making permutation problems into combination problems.

## Exercise Set 13.2

1. Compute $C(4, 3)$.
2. How many different selections can we make of five flags from among twelve different flags?
3. How many two-piece outfits can Charles select from a wardrobe of four jackets and six pairs of pants?
4. How many three-piece suits can Simon select from a wardrobe consisting of four jackets, three vests, and six pairs of pants?
5. If five coins are tossed, in how many different patterns can they land heads or tails?
6. If four dies are tossed, in how many different ways can they land?
7. Ford offers automobiles, with or without automatic transmission, in six different body styles, four different interior fabrics, and nine different colors. How many models would a dealer have to stock in order to display all possible combinations?
8. The Braille system of writing for the blind is based on raised dots in six possible positions. How many different characters are there possible in this system?
9. How many different signals can we form by hoisting any five of five different kinds of flags in a line to the masthead of a ship?
10. How many different selections can there be made of one or more books from a shelf of sixteen books?

## 13.3 Theory of Probability

In everyday language we call something "probable" if we believe it likely to happen, "improbable" if we believe it unlikely to happen, or "certain" if we believe it sure to happen. In **the mathematical theory of probability** we try to define such concepts more precisely so as to assign them measures, or numerical indices, which can be computed, and therefore compared, arithmetically.

As might be expected from its title, the theory deals with the likelihood of things which happen by chance or are selected at random. By definition, these are events which are not influenced to happen one way rather than another by known causes. Simple instances are: the landing of a tossed coin heads or tails, or the selection of "any card" from a shuffled deck.

At first approach, this sort of subject matter may seem to upset all our previous ideas of what mathematics is about. Previously in this subject we have not made such statements, for instance, as "the base angles of an isosceles triangle are probably equal."

When we understand the nature of mathematics, however, we realize that the more precise statements usually made about such matters are just short forms of "If . . . then . . ." hypothetical generalizations. What we really mean, when we take the trouble to repeat it all over again each time, is that "if a triangle is isosceles, then its base angles are equal." These hypothetical generalizations follow from the theoretical assumptions upon which they are based even though there may be no actual physical structure in the entire material universe which is exactly isosceles or which has an angle of precisely 30°.

The corresponding theoretical assumption by which we can put even uncertainties on an exact mathematical basis is that of something happening in a definite number of equally likely ways. Instances are: the assumption of a "well-balanced" coin being equally likely to land with either heads or tails face up, or the assumption of an "unloaded" die being equally likely to stop rolling with any one of its six faces up.

### Basic Definitions

If on any given trial, an event $E$ can happen in $h$ ways, and fail to happen in $f$ ways, out of $w = h + f$ equally likely ways, then the mathematical measure of the probability of $E$ happening is by definition

$p = h/w$, or $h$ chances out of $w$

and the mathematical measure of the probability of $E$ failing to happen (or of the improbability of $E$ happening) is, also by definition

$q = f/w$, or $f$ chances out of $w$.

In other words, $p$ and $q$ are simply numerical indices assigned to the likelihood and unlikelihood, respectively, that $E$ will happen under the assumed theoretical conditions.

From these definitions it follows at once that, in the special case when $h = 0$, then $h/w = 0/w = 0$, and

$p = 0$, the measure of **impossibility**

Also, in the special case when $h = w$, then $h/w = w/w = 1$, and

$p = 1$, the measure of certainty

For instance, let our trial be the tossing of a "well-balanced coin"—by assumption, a coin which can land heads or tails with equal likelihood but cannot stand on edge. And let three events be defined as follows:

$E$—the coin lands heads

$E_1$—the coin stands on edge

$E_2$—the coin lands heads or tails

Then for event $E$, that the coins land heads

$$h = 1, f = 1, w = h + f$$
$$= 1 + 1 = 2$$

and we obtain the probability measures

$$p = h/w = \frac{1}{2} \text{ for } E \text{ happening}$$

$$q = f/w = \frac{1}{2} \text{ for } E \text{ failing to happen}$$

These express numerically the fact that the coin is equally likely to land heads or not to land heads.

Likewise for event $E_1$, that the coin stands on edge

$$h_1 = 0, f_1 = 2$$

$$w_1 = h_1 + f_1 = 0 + 2 = 2$$

and we obtain the probability measures

$$p_1 = h_1/w_1 = \frac{1}{2} = 0 \text{ for } E_1 \text{ happening}$$

$$q_1 = f_1/w_1 = \frac{2}{2} = 1 \text{ for } E_1 \text{ failing}$$

These express numerically the fact that the coin cannot stand on edge.

Finally, for event $E_2$

$$h_2 = 2, f_2 = 0$$

$$w_2 = h_2 + f_2 = 2 + 0 = 2$$

and we obtain the probability measures

$$p_2 = h_2/w_2 = \frac{2}{2} = 1 \text{ for } E_2 \text{ happening}$$

$$q_2 = f_2/w_2 = \frac{0}{2} = 0 \text{ for } E_2 \text{ failing}$$

These express numerically the fact that the coin must certainly land either heads or tails (not heads) under the given assumptions.

From the above definitions it is also clear that the probability measures $p$ and $q$ must always have values from 0 to 1 inclusive:

$$0 \le p \le 1, \text{ and } 0 \le q \le 1$$

This is only reasonable since we should not expect a measure of probability to be less than that of impossibility (0) or greater than that of certainty (1).

Finally, since $w = h + f$ by definition, we see that

$$\frac{h}{w} + \frac{f}{w} = \frac{h + f}{w} = \frac{w}{w} = 1$$

Hence, by substitution of $p = h/w$, etc.,

$$p + q = 1, p = 1 - q, q = 1 - p$$

Any one of these last three relationships may serve as a partial check upon the accuracy of work in calculating $p$ and $q$. (The "check" works unless pairs of errors have been made exactly compensating for each other.) The last two relationships may also serve for finding either $p$ or $q$ once the other has been computed.

In the preceding illustration of event $E$, for instance;

$$p + q = \frac{1}{2} + \frac{1}{2} = 1 \text{ Partial check}$$

Also, if we had not separately computed $q$, we could have found it by substitution, as

$$q = 1 - p = 1 - \frac{1}{2} = \frac{1}{2} \text{ (as before)}$$

Values of the probability fractions $p$ and $q$ may also be expressed in decimal or in percentage form. In the latter case they are called percentage chances. From the above illustration again, for instance:

$$p = \frac{1}{2} = 0.5, \text{ or a } 50\% \text{ chance of } E$$
happening

$$q = \frac{1}{2} = 0.5, \text{ or a } 50\% \text{ chance of } E$$
failing

In this special case, the "chances" may be said to be "50-50"; or, if betting is involved, it can be said to be based on "even money" terms.

A different, but equivalent, way of expressing the ideas of probability and improbability is in terms of "odds." By definition, the ratio

$$h{:}f, \text{ or } h \text{ to } f$$

states the odds for $E$ happening; and the inverse ratio

$$f{:}h, \text{ or } f \text{ to } h$$

states the odds against $E$ happening. In the case of the preceding illustration, for instance, the odds are:

$$h{:}f, \text{ or } 1 \text{ to } 1 \text{ for } E \text{ happening, and}$$

$$f{:}h, \text{ or } 1 \text{ to } 1 \text{ against } E \text{ happening}$$

In this special case where both ratios are the same we say "the odds are even."

When the odds for $E$ happening are greater than $1{:}1$, they may also be called the odds in favor of $E$.

If the statement of a probability problem is such that you can determine $h$, $f$, and $w$ by actual count, you can always solve the problem directly by the method of the above illustrations.

EXAMPLE 1: If two well-balanced coins are tossed, what are the probabilities of, and the odds on, the following events:

$E_1$—both land heads

$E_2$—lands heads and one tails

$E_3$—neither lands heads

$E$—at least one lands heads

SOLUTION: Designate the respective heads and tails sides of the two coins in some arbitrary order as $H_1$, $T_1$, and $H_2$, $T_2$. Then the assumption that the coins are well balanced means that they are equally likely to land with any one of the four possible intergroup combinations of their faces up.

$$H_1H_2, \quad H_1T_2, \quad T_1H_2, \quad T_1T_2$$

For event $E_1$, then, $w_1 = 4$, $h_1 = 1$ (the count of the $H_1H_2$ possibility), and the probability of $E_1$ happening is

$$p_1 = h_1/w_1 = \frac{1}{4}, \text{ or } 1 \text{ chance out of } 4$$

which may also be expressed as

$$p_1 = 0.25, \text{ or a } 25\% \text{ chance}$$

Alternatively, since $f_1 = w_1 - h_1 = 4 - 1 = 3$, the odds on $E_1$ happening are:

$$h_1{:}f_1 = 1{:}3, \text{ or } 1 \text{ to } 3 \text{ for, and}$$

$$f_1{:}h_1 = 3{:}1, \text{ or } 3 \text{ to } 1 \text{ against}$$

For event $E_2$ next, $w_2 = 4$ as before, but $h_2 = 2$ (the count of the $H_1T_2$ and $T_1H_2$ possibilities). Hence, the probability of $E_2$ happening is:

$$p_2 = h_2/w_2$$
$$= \frac{2}{4}, \text{ or } 2 \text{ chances out of } 4$$

Equivalently, by reduction of the fraction to lowest terms, etc., this is:

$$p_2 = \frac{1}{2}, \text{ or } 1 \text{ chance out of } 2$$

$$= 0.5, \text{ or a } 50\% \text{ chance}$$

And since $f_2 = w_2 - h_2 = 4 - 2 = 2$, the odds on $E_2$ happening are:

$$h_2 : f_2 = 2 : 2$$
$$= 1 : 1, \text{ or even (for or against)}$$

For event $E_3$, again $w_3 = 4$, but (as for $E_1$) $h_3 = 1$ (this time the count of the $T_1T_2$ possibility). Hence, the probability of, and the odds on, $E_3$ happening are exactly the same as for $E_1$ happening.

For event $E$, however, although $w = $ four again, now $h = 3$ (the count of the $H_1H_2$, $H_1T_2$, and $T_1H_2$ possibilities).

Hence, the probability of $E$ happening is

$$p = h/w = \frac{3}{4}, \text{ or 3 chances out of 4}$$

$$= 0.75, \text{ or a 75\% chance}$$

Alternatively, since $f = w - h = 4 - 3 = 1$, the odds on $E$ happening are:

$$h : f = 3 : 1, \text{ or 3 to 1 for, and}$$

$$f : h = 1 : 3, \text{ or 1 to 3 against}$$

EXAMPLE 2: A deck of bridge cards is shuffled and split at random. What is the probability that the card thus exposed is: (a) of a red suit—hearts or diamonds? (b) a diamond? (c) a queen? (d) the queen of diamonds?

SOLUTION: Since there are $w = $ fifty-two cards in the deck and $h_a = $ twenty-six of these are of the two red suits, the probability of the exposed card having a red face is:

**a)** $p_a = h_a/w = \dfrac{26}{52} = \dfrac{1}{2}$

Since $h_b = 13$ of the cards are diamonds, the probability of the exposed card being a diamond is

**b)** $p_b = h_b/w = \dfrac{13}{52} = \dfrac{1}{4}$

Since $h_c = 4$ of the cards are queens, the probability of the exposed card being a queen is:

$$p_c = h_c/w = \frac{4}{52} = \frac{1}{13}$$

But since there is only $h_d = 1$ queen of diamonds in the deck, the probability of the exposed card being this one is

$$p_d = h_d/w = \frac{1}{52}$$

## Relative Probabilities

If the probabilities of events $E_1$ and $E_2$ are $p_1$ and $p_2$ respectively, then by definition event $E_1$ is $p_1/p_2$ **times as probable** (likely to happen) as event $E_2$. From the findings in the solution of EXAMPLE 2, for instance, we may say that when a card is drawn at random from a shuffled bridge deck, the event of its being a diamond is $(^{13}\!/\!_{52}) \div (^4\!/\!_{52})$ $= ^{13}\!/\!_4 = 3\frac{1}{4}$ times as probable as the event of its being a queen; or, more briefly, such a card is $3\frac{1}{4}$ times more likely to be a diamond than it is to be a queen (although, of course, it may be both in the special case of the queen of diamonds).

To illustrate the relative likelihoods of all the possible outcomes of a game, it is sometimes convenient to prepare a table of all its basic events and their consequent probabilities.

In the game of dice, for instance, each of two dice has its faces numbered from one to six by a corresponding number of spots, and when the dice come to rest after being rolled, the players add the two of these numbers faced upward. We have already seen that the total number of possible combinations of the faces of two dice is $6^2 = 36$. But the corresponding game events (sums of two numbers each from one to six can also be catalogued in a table constructed as follows: all the possible face numbers from one to six for an arbitrarily selected first die are written on successive lines down the left side; the same numbers for the other die are written at the heads of columns along the top; and the thirty-six different possible sums of one number from each are written at corresponding intersections of lines and columns, thus:

(Face numbers of second die)

| + | 1 | 2 | 3 | 4 | 5 | 6 |
|---|---|---|---|---|---|---|
| 1 | 2 | 3 | 4 | 5 | 6 | 7 |
| 2 | 3 | 4 | 5 | 6 | 7 | 8 |
| 3 | 4 | 5 | 6 | 7 | 8 | 9 |
| 4 | 5 | 6 | 7 | 8 | 9 | 10 |
| 5 | 6 | 7 | 8 | 9 | 10 | 11 |
| 6 | 7 | 8 | 9 | 10 | 11 | 12 |

(Face numbers of first die)

(Sums of both face numbers)

By counting identical entries in the body of this table, we can now see at a glance in how many ways, out of $w = 36$, a pair of die faces can add up to each possible total. Hence, assuming that all $w = 36$ ways are equally likely—that the dice are not "loaded," that is—to compute the corresponding probabilities we have only to substitute these values of $h$ in the definition formulas.

EXAMPLE 3: What is the sum (of two face numbers) most likely to be rolled on a pair of dice? How does the probability of this sum being rolled compare with the probabilities of others?

SOLUTION: Let $h_1, h_2, \ldots h_{13}$ and $p_1, p_2, \ldots p_{13}$ be respectively the number of favorable happenings and the probabilities for the sums, 1, 2, . . . 13. Then by inspection of the above table and actual count we find:

$$h_1 = h_{13} = 0, \quad h_2 = h_{12} = 1$$
$$h_3 = h_{11} = 2, \quad h_4 = h_{10} = 3, \text{etc.}$$

And:

$$p_1 = p_{13} = \frac{0}{36} \text{ (or 0)}$$
$$p_2 = p_{12} = \frac{1}{36}$$
$$p_3 = p_{11} = \frac{2}{36}\left(\text{or }\frac{1}{18}\right)$$
$$p_4 = p_{10} = \frac{3}{36}\left(\text{or }\frac{1}{12}\right)$$
$$p_5 = p_9 = \frac{4}{36}\left(\text{or }\frac{1}{9}\right)$$
$$p_6 = p_8 = \frac{5}{36}$$
$$p_7 = \frac{6}{36}\left(\text{or }\frac{1}{6}\right)$$

Thus we see that the most likely number to be rolled is seven, with a probability of six chances out of thirty-six, or of one out of six. By comparison, six and eight are only $(\frac{5}{36})/(\frac{6}{36}) = \frac{5}{6}$ as likely to be rolled; five and nine are only $\frac{4}{6} = \frac{2}{3}$ as likely; four and ten are only $\frac{3}{6} = \frac{1}{2}$ as likely; three and eleven are only $\frac{2}{6} = \frac{1}{3}$ as likely; two and twelve are only $\frac{1}{6}$ as likely; and one and thirteen are impossible (as are all numbers higher than 13).

EXAMPLE 4: On his first cast a dice player rolls a ten, which he regards as his "lucky number." To win, he must again roll this number, called "his point," before he rolls a seven. Another player now offers a "side bet" on "even money" terms that the first player will not make his point. Is this a fair wager?

SOLUTION: Definitely not! As we have already seen from the preceding solution, a ten is only half as likely to be rolled as a seven. Hence, a fair bet would require the holder of the dice to be given 2-to-1 odds. If he takes the bet on the superstitious theory that ten is his "lucky number," then so much the worse for him in the long run of several trials!

---

**Warning**

If event $E_1$ is $h_1/h_2$ times as probable as event $E_2$, this means only that the **odds** are $h_1/h_2$ for $E_1$ happening before $E_2$. To compute the **probability** of event $E$, that $E_1$ will happen before $E_2$, we must recognize that $E$ can happen in only $h = h_1$ out of $w = h_1 + h_2$ possible ways.

---

Hence the probability that $E_1$ will happen before $E_2$ is only

$$p = \frac{h_1}{h_1 + h_2}$$

In the preceding case, for instance, the probability that a four will be rolled before a seven is not $\frac{3}{6} = \frac{1}{2}$, *but* . . .

$$p = \frac{3}{3 + 6} = \frac{3}{9} = \frac{1}{3}$$

or equivalently

$$p = \frac{1}{1 + 2} = \frac{1}{3}$$

Likewise, the probability of a seven being rolled before a four is

$$q = \frac{2}{2 + 1} = \frac{2}{3}$$

Partial Check:

$$p + q = \frac{1}{3} + \frac{2}{3} = \frac{3}{3} = 1$$

## Addition of Probabilities

Any two events are by definition **mutually exclusive** if the happening of either excludes the possible happening of the other on the same trial.

For instance, the event $e$ of a coin landing heads is mutually exclusive with the event $e'$ of the same coin landing tails on the same trial (toss). Similarly, the eleven events $E_2, E_3 . . . E_{12}$, of a pair of dice rolling sums of 2, 3 . . . 12, respectively, are all mutually exclusive with respect to each other on the same trial (roll).

In other words, mutually exclusive events are alternative possible outcomes of the same trial.

From this definition, it follows that if $E_1$ and $E_2$ are any two mutually exclusive events, then $E_1$ can happen in $h_1$ ways and $E_2$ in $h_2$ ways out of the same $w$ equally likely ways, but the $h_1$ ways of $E_1$ happening must all be different from the $h_2$ ways of $E_2$ happening.

Suppose, then, that $E$ is the event of any one of $n$ mutually exclusive events, $E_1$, or $E_2$, or $E_3$ or . . . $E_n$. Then $E$ can happen in any of the ways;

$$h = h_1 + h_2 + \text{.} \ . \ . + h_n$$

out of the same $w$ equally likely ways. Therefore:

$$\frac{h}{w} = \frac{h_1 + h_2 + \ . \ . \ . h_n}{w}$$

$$= \frac{h_1}{w} + \frac{h_2}{w} + \cdots + \frac{h_n}{w}$$

But, by definition:

$$p = \frac{h}{w}, p_1 = \frac{h_1}{w} \cdots p_n = \frac{h_n}{w}$$

Therefore, by substitution:

$$p = p_1 + p_2 + \cdots p_n$$

---

If $E_1$, $E_2 \cdots E_n$ are $n$ mutually exclusive events, then the probability of $E_1$ or $E_2$ or $\cdots$ $E_n$ happening is equal to the sum of the probabilities of each of these events happening separately.

---

In EXAMPLE 1 above, for instance, the events $E_1$ (that two tossed coins land heads) and $E_2$ (that one lands heads and one tails) are mutually exclusive. Also, $E$ (that at least one lands heads) is the event that either $E_1$ or $E_2$ happens. In the direct solution of that problem we found, moreover, that

$$p_1 = \frac{1}{4}, p_2 = \frac{1}{2}, p = \frac{3}{4}$$

This solution is now confirmed by the above formula, as follows:

$$p_1 + p_2 = \frac{1}{4} + \frac{1}{2} = \frac{3}{4}$$
$$= p, \text{ as anticipated}$$

In the special case where $E_1$, $E_2 \cdots E_n$ are all the mutually exclusive events which can possibly happen in the given $w$ equally likely outcomes of the same trial, then by the same reasoning:

$$p = p_1 + p_2 + \cdots p_n = 1$$

From EXAMPLE 1 again, for instance, we also have the event $E_3$ (that two tossed coins both land tails). With $E_1$ and $E_2$, this completes the mutually exclusive possibilities for the outcome of the trial of tossing two coins. Also, $p_3 = \frac{1}{4}$, and

$$p = \frac{1}{4} + \frac{1}{2} + \frac{1}{4} = 1, \text{ as anticipated}$$

Or, from EXAMPLE 3 concerning cast dice:

$$p = \left( \frac{\begin{array}{c} 1 + 2 + 3 + 4 + 5 + 6 + \\ 5 + 4 + 3 + 2 + 1 \end{array}}{36} \right)$$

$$= \frac{36}{36} = 1, \text{ as anticipated}$$

The above probability **addition formulas** can often be applied to shorten the computation of probabilities, or to find further probabilities in terms of those already computed.

EXAMPLE 5: Using the data in the solution of EXAMPLE 3, find the probability that a player will roll a seven or an eleven on his first cast of the dice.

SOLUTION: The probability is:

$$p = p_7 + p_{11} = \frac{6}{36} + \frac{2}{36} = \frac{8}{36}$$

$$= \frac{2}{9} \text{ or 2 chances out of 9}$$

EXAMPLE 6: Find the corresponding probability that he will roll a two or a three or a twelve on his first cast.

SOLUTION: The probability is:

$$p = p_2 + p_3 + p_{12} = \frac{1}{36} + \frac{2}{36} + \frac{1}{36}$$

$$= \frac{4}{36} = \frac{1}{9}, \text{ or 1 chance out of 9}$$

Two events which are not mutually exclusive may be so because they are (partially) overlapping. In EXAMPLE 2 above, for instance, the event $E_b$ (that a card drawn from a deck at random be a diamond) and the event $E_c$ (that the card be a queen) are not mutually exclusive, but are (partially) overlapping in the case of the event $E_d$ (that the card be the queen of diamonds).

If $E_1$ and $E_2$ are any two (partially) overlapping events with probabilities $p_1$ and $p_2$ respectively, and if $p(E_1$ and $E_2)$ is the probability that both $E_1$ and $E_2$ happen, then the probability that either $E_1$ or $E_2$ happens is:

$$p(E_1 \text{ or } E_2) = p_1 + p_2 - p(E_1 \text{ and } E_2)$$

EXAMPLE 7: What is the probability that a card drawn at random from a bridge deck is either a diamond or a queen?

SOLUTION: From the solution of EXAMPLE 2 we already know that the probability of such a card's being a diamond is $13/52 = 1/4$, that the probability of its being a queen is $4/52 = 1/13$, and that the probability of its being both a queen and a diamond is $1/52$. Hence, the probability of its being either a queen or a diamond is:

$$p = \frac{13}{52} + \frac{4}{52} - \frac{1}{52}$$

$$= \frac{13 + 4 - 1}{52}$$

$$= \frac{16}{52} = \frac{4}{13}$$

EXAMPLE 8: The probability that a poker hand be "any straight" (including a "straight flush") is $p_1 = 0.003940$. The probability that it be "any flush" (including a "straight flush") is $p_2 = 0.001970$. The probability that it be a "straight flush" is $p_{12} = 0.000014$. What is the probability that a poker hand be either a "straight" or a "flush"?

SOLUTION: As in the preceding solution:

$$\begin{aligned}
p_1 &= \phantom{+}0.003940 \\
+p_2 &= +0.001970 \\
\hline
p_1 + p_2 &= \phantom{+}0.005910 \\
-p_{12} &= -0.000014 \\
\hline
p &= \phantom{+}0.005896
\end{aligned}$$

## Multiplication of Probabilities

Events which are neither mutually exclusive nor (partially) overlapping are separate events.

By definition, any two events $e_1$ and $e_2$ are separate if the set of $w_1$ ways in which $e_1$ can happen or fail to happen are completely distinct from the $w_2$ ways in which $e_2$ can happen or fail to happen. For instance, the event $e_1$ that a first coin land heads is separate from the event $e_2$ that a (different) second coin land heads, because the $w_1$ = two ways in which the first can land either heads or tails are completely distinct from the $w_2$ = two ways in which the second coin can land either heads or tails.

Also by definition, if $e_1$, $e_2 \cdots e_n$ are $n$ separate events and $E$ is the event that $e_1$, and $e_2$, $\cdots$ and $e_n$ all happen (either concurrently or in succession) on the same trial, then $E$ is a multiple event with $e_1$, $e_2 \cdots e_n$ as its (separate) constituent events. For instance, the event $E$ (that two coins land heads) is a multiple event consisting of the two separate constituent events, $e_1$ (that the first coin lands heads) and $e_2$ (that the second coin lands heads).

Now let $E$ be the multiple event that $n$ separate constituent events, $e_1$ and $e_2$, and $\cdots e_n$ all happen on the same trial, and let $h_1$ be the number of ways in which $e_1$ can happen out of a total of $w_1$ possible ways distinct from the $w_2$, $w_3 \cdots w_n$ distinct ways of $e_2$, $e_3 \cdots e_n$ happening or failing to happen, etc. Then, by intergroup combinations of the possibilities, $E$ can happen in

$$h = h_1 h_2 \cdots h_n \text{ ways}$$

out of a total of

$$w = w_1 w_2 \cdots w_n \text{ ways}$$

Therefore:

$$\frac{h}{w} = \frac{h_1 h_2 \cdots h_n}{w_1 w_2 \cdots w_n} = \frac{h_1}{w_1} \cdot \frac{h_2}{w_2} \cdots \frac{h_n}{w_n}$$

And so, if all these ways are equally likely, then by substitution of $p = h/w$, etc.:

$$p = p_1 p_2 \cdots p_n$$

The probability of the happening of a multiple event on any given trial is equal to the product of separate probabilities of its $n$ separate constituent events.

Along with the preceding addition formulas for the probabilities of mutually exclusive events ($E_1$ or $E_2$ or $\cdots E_n$), this multiplication formula for the probabilities of separate events ($e_1$ and $e_2$ and $\cdots e_n$) can shorten the work of computing many probabilities.

EXAMPLE 9: What are the odds against a coin landing heads eight times in a row?

SOLUTION: The landings of the same coin eight different times (or of different coins at the same time) are separate trials. Moreover, for each the probability of the coin landing heads is ½ as we have already seen. Hence, the probability of the multiple event of the coin landing heads all eight times is, by the multiplication theorem:

$$p = \left(\frac{1}{2}\right)^8 = \frac{1}{256}$$

Accordingly:

$$h = 1, w = 256$$

$$f = w - h = 256 - 1 = 255$$

and the odds on the multiple event are

$f:h = 255$ to 1 against its happening

EXAMPLE 10: A coin has been tossed and fallen heads seven times in a row. An excited spectator, aware of the result of the previous solution, offers to bet 100 to 1 that it will not come heads again! What is the wisdom of his wager?

SOLUTION: The man is a fool who completely misses the point about separate events. His bet would have been a very advantageous one if made before the series of tosses began. But the solution of EXAMPLE 7 is valid only because the separate probability of each separate constituent event is exactly 1 chance out of 2. And this holds for the eighth trial, the first trial, or the millionth trial, considered separately from the series in which it occurs.

Two separate events are said to be **independent** if the happening of neither affects the probability of the happening of the other. But two separate events are said to be **dependent** if the happening of either does affect the probability of the happening of the other. Numerical consequences of this distinction are illustrated in the following two examples.

EXAMPLE 11: One card is drawn from each of two different decks (or else a card is drawn from one deck and returned to it after which another card is drawn from the same deck). What is the probability that both cards are aces?

SOLUTION: The two separate events, $E_1$ and $E_2$, of drawing an ace are here independent because the probability of each

$$p_1 = p_2 = \frac{4}{52} = \frac{1}{13}$$

is not affected by the occurrence of the other. Therefore, the probability of both cards being aces is

$$p = p_1 p_2 = \frac{1}{13}\frac{1}{13} = \frac{1}{169}$$

EXAMPLE 12: Two cards are drawn in succession from the same deck without the first being returned to it. What is the probability that both are aces?

SOLUTION: As before, the probability of event $E_1$, that the first card be an ace, is $p_1 = \frac{1}{13}$. But now the event $E_3$, that the second card also be an ace, is dependent upon event $E_1$ having happened. For if an ace has already been drawn from the deck, then there are only $h_3 = 3$ aces left in a "short deck": of $w_3 = 51$ cards. Therefore, the probability of $E_3$ is

$$p_3 = h_3/w_3 = \frac{3}{51} = \frac{1}{17}$$

and therefore the probability of both cards being aces is now only

$$p = p_1 p_3 = \frac{1}{13} \frac{1}{17} = \frac{1}{221}$$

## Misconceptions and Superstitions

The uncertainties of chance, which is the subject matter of probability theory, is also the object of many hopes and fears. Understandably, therefore, the field is one in which wishful thinking often leads to misconceptions, or to superstitions based on misconceptions.

Two simple instances have already been pointed out in EXAMPLES 4 and 10 above. Another now follows.

EXAMPLE 13: By the rules in the game of dice, the player who rolls the dice must match bets placed against him on an "even money" basis. He can win either (a) by rolling a "natural"—seven or eleven—on his first cast; or else (b) by rolling a "point"— four, five, six, eight, nine, or ten—on his first cast and then "making his point" by rolling the same number again before a seven. On the other hand, he can lose either (c) by rolling "craps"—two, three, or twelve—on his first cast; or else (d) by rolling seven before his "point."

Against the background of these rules, amateur dice players are known to "feel" that it is "luckier" to roll the dice themselves, even when they have no question about the honesty of their fellow players. But professional gambling-house operators always require the "customer" to roll the dice rather than their own employee. Which policy is mathematically more advantageous?

SOLUTION: By applying the probability addition and multiplication formulas to the data of EXAMPLES 3 and 4 above, we can now compute the probabilities of all the possible outcomes of this game—both favorable and unfavorable to the one who cast the dice. For instance, in the case with a $\frac{1}{12}$ probability that the player rolls a four as his point, we know that he has only one/ one + two = one/three chance thereafter of making this point. Hence, the probability of his winning in this instance is the product of the probabilities of two separate constituent events:

$$\frac{1}{12} \cdot \frac{1}{3} = \frac{1}{36} = 2.77\%$$

Moreover, his probability of winning in the mutually exclusive event of his rolling a ten as his point is the same. Hence, the probability of his winning in either of these cases is the sum of the probabilities of these two mutually exclusive events:

$$\frac{1}{36} + \frac{1}{36} = 2 \cdot \frac{1}{36} = \frac{1}{18} = 5.55\%$$

On the other hand, the probability of his losing in the cases of these same two mutually exclusive "point" possibilities is the corresponding sum of two products:

$$2 \cdot \frac{1}{12} \cdot \frac{2}{2+1} = \frac{4}{36} = \frac{1}{9} = 11.11\%$$

Analyzing all other possibilities in the same way, we obtain the following table. There we see that the one who rolls the dice has a 49.29% chance of winning as opposed to a 50.71% chance of losing. This means that, for a fair game, turns to roll the dice should be alternated among all the players. It also means that, even if there is no "admission fee" or "percentage charge for cashing chips," the professional gambling house always wins in the long run—for, in this case, the "house" literally has the customer who rolls the dice working for it!

**TABLE OF PROBABILITIES FOR GAME EVENTS IN DICE**

| POSSIBLE CASTS BY PLAYER | PROBABILITY CALCULATIONS | PERCENTAGE CHANCES | |
|---|---|---|---|
| | | TO WIN | TO LOSE |
| "Craps": 2, 3, or 12 | $\dfrac{1+2+1}{36} = \dfrac{4}{36} = \dfrac{1}{9} =$ | | 11.111 |
| "Point" 4 or 10, made | $2 \cdot \dfrac{1}{12} \cdot \dfrac{1}{1+2} = \dfrac{2}{36} = \dfrac{1}{18} =$ | 5.555 | |
| "Point" 4 or 10, lost | $2 \cdot \dfrac{1}{12} \cdot \dfrac{2}{2+1} = \dfrac{4}{36} = \dfrac{1}{9} =$ | | 11.111 |
| "Point" 5 or 9, made | $2 \cdot \dfrac{1}{9} \cdot \dfrac{4}{4+6} = \dfrac{8}{90} = \dfrac{4}{45} =$ | 8.889 | |
| "Point" 5 or 9, lost | $2 \cdot \dfrac{1}{9} \cdot \dfrac{6}{6+4} = \dfrac{12}{90} = \dfrac{2}{15} =$ | | 13.333 |
| "Point" 6 or 8, made | $2 \cdot \dfrac{5}{36} \cdot \dfrac{5}{5+6} = \dfrac{50}{396} = \dfrac{25}{198} =$ | 12.626 | |
| "Point" 6 or 8, lost | $2 \cdot \dfrac{5}{36} \cdot \dfrac{6}{6+5} = \dfrac{60}{396} = \dfrac{5}{33} =$ | | 15.151 |
| A "natural": 7 or 11 | $\dfrac{6+2}{36} = \dfrac{8}{36} = \dfrac{2}{9} =$ | 22.222 | |
| | **Total percentage chances:** | 49.29 | 50.71 |
| | **Partial Check: 49.29% + 50.71% =** | 100% | |

## Mathematical Expectation

As distinguished from psychological or subjective expectation, **mathematical expectation** is defined as the product, $p \cdot A$, of the probability $p$ that a particular event will happen, and the amount $A$ one will receive if it does happen.

Suppose, for instance, that a dice player has bet $5.00. When this amount is "matched," it results in a "pot" of $A = \$10$ which the player will receive if he wins. From the preceding table we know that the probability of his winning is $p = 49.29\%$. Hence, his mathematical expectation is:

$$p \cdot A = 0.4929(\$10) = \$4.93$$

to the nearest cent. However, if he should roll a four or a ten as his "point," then the probability of his winning drops to $p = \frac{1}{3}$ and his mathematical expectation thereafter is only

$$\$10/3 = \$3.33$$

to the nearest cent, whereas that of the "house" then increases from $5.07 to $6.67 to the nearest cent.

EXAMPLE 14: In a lottery, 20,000 tickets are to be sold at 25¢ each, and the one prize is $2,500. In another lottery, 200,000 tickets are to be sold at $1.00 each, and the prizes are one of $50,000 and 25 of $1,000. Mr. Sloe is thinking of making a $1.00 "investment" in either. He is tempted by the first lottery because he thinks it will give him "four times as many chances (by which he means tickets) for the same money." But he is also tempted by the second lottery because he thinks "there is so much more prize money." Compare his actual mathematical expectations in the two cases.

SOLUTION: In the first lottery, Sloe would have a $4/20,000 = 0.0002$ probability of winning $2,500. Hence, for his $1.00 cost of four tickets he would receive a mathematical expectation worth
$$V_1 = 0.0002(\$2,500) = 50¢$$

In the second lottery he would have a $1/200,000 = 0.000005$ probability of winning $50,000, plus a $25/200,000 = 0.000125$ probability of winning $1,000. Hence for the same $1.00 cost he would receive a mathematical expectation worth

$$V_2 = 0.000005(\$50,000)$$
$$+ 0.000125(\$1,000)$$
$$= 25¢ + 12\frac{1}{2}¢ = 37\frac{1}{2}¢$$

Consequently, we might advise Sloe that neither purchae is an "investment," and the second is by 25% an even worse "speculation" than the first.

## Applying Combinations Formulas

In the probability problems considered thus far it has been possible to count values of $h, f,$ and $w$ directly. In more complicated problems, however, these quantities are better computed by formula as in the preceding chapter.

EXAMPLE 15: What are the odds against the event of a poker player being dealt a two-pair hand—defined as one having a pair in each of two different ranks plus a fifth card in a still different rank?

SOLUTION: From EXAMPLE 4 we know that a poker hand can be dealt in a total of $w = 2,598,960$ different ways. To compute $h$ we can now reason that two different ranks may be combined in $C(13, 2)$ different ways; that in each of these cases 2 pairs of different ranks can occur in $C(4, 2)$ times $C(4, 2) = 6 \cdot 6 = 36$ different ways; and that a fifth card from a still different rank can occur in 44 different ways, for a product of

$$h = C_2^{13}C_2^4C_2^444 = 78 \cdot 36 \cdot 44$$
$$= 123,552$$

Therefore:
$$f = w - h = 2,598,960 - 123,552$$
$$= 2,475,408$$

and the odds against dealing a two-pair hand are

$$f{:}h = 2,475,408{:}123,552 = 20 \text{ to } 1$$

rounded off to the nearest whole number.

**TABLE OF MATHEMATICAL ODDS FOR HANDS DRAWN IN POKER**

| HAND | COMPUTATION OF h | VALUE OF h | ODDS AGAINST |
|------|------------------|-----------:|-------------:|
| Straight flush | $10C_1^4 = 10 \cdot 4 =$ | 40 | 64,973 to 1 |
| Four of a kind | $C^{13}C_4^4C_1^{12}C_1^4 = 13 \cdot 1 \cdot 12 \cdot 4 =$ | 624 | 4,164 to 1 |
| Full house | $C_2^{13}C_2^4C_3^4 = 78 \cdot 2 \cdot 6 \cdot 4 =$ | 3,744 | 693 to 1 |
| Flush (non-straight) | $C_5^{13}C_1^4 - 40 = 5,148 - 40 =$ | 5,108 | 508 to 1 |
| Straight (non-flush) | $10(4^5) - 40 = 10,240 - 40 =$ | 10,200 | 254 to 1 |
| Three of a kind | $C_1^{13}C_3^4C_2^{12}4^2 = 13 \cdot 4 \cdot 66 \cdot 16 =$ | 54,912 | 46 to 1 |
| Two pairs | $C_2^{13}C_2^4C_2^444 = 78 \cdot 6 \cdot 6 \cdot 44 =$ | 123,552 | 20 to 1 |
| One pair | $C_1^{13}C_2^4C_2^{12}4^3 = 13 \cdot 6 \cdot 220 \cdot 64 =$ | 1,098,240 | 1.4 to 1 |
| Other | $w - \text{(all the above)} =$ | 1,302,540 | 1 to 1 |
| Any five cards | $w = C_5^{52} = \dfrac{52!}{5!47!} =$ | 2,598,960 | 0.0 to 1 |

EXAMPLE 16: What is the probability of dealing (a) a bridge hand consisting of thirteen cards all of the same suit? (b) four such hands from the same deck? (c) a bridge hand with a four, four, three, two suit distribution?

SOLUTION: (a) One bridge hand of thirteen cards can be dealt from a fifty-two-card deck in any of

$$w_a = C_{13}^{52} = \frac{52!}{13!39!}$$
$$= 635,013,559,600 \text{ ways}$$

Of these, only $h_a = 4$ have all thirteen cards in the same suit. Hence, the probability of such a deal is

a) $p_a = 4/w_a = 1/158,753,389,900$

Since it makes no difference whether each of four hands is dealt, from a shuffled deck, one card at a time or all thirteen cards at a time, let us for convenience assume the deal to be the latter. Then the four hands can be dealt consecutively to four players in

$$C_{13}^{52}, C_{13}^{39}, C_{13}^{26}, \text{ and } C_{13}^{13}, \text{ ways}$$

respectively, and the product of these four quantities would give the number of ways in which bridge hands can be permuted among four players. Dividing this product by factorial 4, therefore, we find the total number of combinations of such hands to be

$$w_b = C_{13}^{52}C_{13}^{39}C_{13}^{26}C_{13}^{13}/4!$$
$$= \frac{52!}{13!39!} \cdot \frac{39!}{13!26!} \cdot \frac{26!}{13!13!} \cdot \frac{13!}{13!} \cdot \frac{1}{4!}$$
$$= \frac{52!}{(13!)^4 4!}$$

when like factors are canceled from numerator and denominator. This last quantity has been machine computed to be 2,235 followed by 24 additional digits. Since $h_b = 1$, therefore, the required probability is

b) $p_b = h_b/w_b$

$$= \frac{1}{2,235 \text{ trillions of trillions}}$$

There are $C(4, 2) = 6$ ways of choosing the two suits in which (the same numbers of) four cards are to be dealt, and $2C_2^2 = 2 \cdot 1 = 2$ different ways of next choosing the two suits in which (the different numbers of) three and two cards are to be dealt. Hence,

$$h_c = 6C_4^{13}C_4^{13}2C_3^{13}C_2^{13}$$

$$= 12 \cdot \frac{13!}{4!9!} \cdot \frac{13!}{4!9!} \cdot \frac{13!}{3!10!}$$

$$\cdot \frac{13!}{2!11!}$$

$$= \frac{(13!)^4}{11!10!(9!)^2(4!)^2}$$

$$= \frac{13^4 12^4 11^3 10^2}{4^2 3^2 2^2}$$

$$= 13^4 12^2 11^3 5^2$$

$$= 136,852,875,100$$

Divided by $w_a$ (rounded off to 635 billion) from above, this gives us the requird probability as

c) $p_c = h_c/w = 0.2155$

or somewhat better than 1 chance in 5.

## "Continuous Probability"

In all preceding parts of this chapter we have thus far considered only events $E$ for which it was possible to count or to compute a definite number of ways $h$ in which $E$ could happen out of an equally definite number of ways $w$ in which $E$ could either happen or fail to happen. These are variously called **discontinuous, arithmetic,** or **finite events.** And, by a somewhat inappropriate transfer of adjectives, the corresponding measures of the likelihoods of their happening are called discontinuous, arithmetic, or finite probabilities.

But suppose we are told that a "stick" is broken "anywhere at random," and we are asked to compute the probability of its being broken closer to its midpoint than to either end. Since the stick can be broken in an infinite (indefinitely large) number of points, we cannot count, or otherwise compute, any definite values for either $h$ or $w$. In such a case, the possibilities are said to be continuous, geometric, or infinite events. And obviously our previously stated definitions of probability and improbability do not, in their original form, apply to such events.

However, for convenience in referring to its points and to segments of its length, let us suppose the "stick" in the preceding instance to be a common 12-inch ruler. Obviously such a stick will be broken closer to the 6-inch mark of its mid-point anywhere between its 3-inch and 9-inch marks. If, temporarily to simplify the problem therefore, we were to consider it breakable only at its full-inch marks, then we should have $w = 11$ (the count of the 1-inch, 2-inch . . . 11-inch marks), and $h = 5$ (the count of the 4-inch, 5-inch . . . 8-inch marks). Hence, in this simplified case,

$$p = h/w = 5/11 = 0.4545 \ldots$$

Or if, coming a little closer to the original problem's condition, we were to consider the ruler to be breakable at any of its eighth-of-an-inch marks, then we should have $w = 95, h = 47$

$$p = h/w = 47/95 = 0.4947 \ldots$$

And by continuing this process of considering the ruler to be breakable in more and more points, we begin to suspect that, as can actually be proven in calculus, the value of $p = h/w$ comes closer and closer to $\frac{1}{2} = 0.5$, which is the ratio of the length of the middle 6 inches of the ruler (between the 3-inch mark and 9-inch marks) to its entire length.

In such a case, therefore, we define the— so-called—**continuous, geometric, or infinite probability** of the continuous, geometric, or infinitely varied event as the limit $p$ to which the ratio $h/w$ comes closer and closer as the number of possible cases increases indefinitely. But in a literally geometric case such as that just considered, we may take this limit to be the ratio of the corresponding geometric lengths, areas, volumes, angles, or even time intervals, involved in the statement of the problem. In the present illustration, for instance,

$$p = \frac{(6 \text{ inches})}{(12 \text{ inches})} = \frac{1}{2}$$

EXAMPLE 17: If a stick of length $L$ is broken anywhere at random, what is the probability that one piece is more than twice as long as the other?

SOLUTION: One piece will be more than twice as long as the other only if the break occurs at a point less than $L/3$ distant from either end—that is, in one of the two segments marked more heavily in the accompanying diagram. Therefore:

$$p = \frac{(L/3 + L/3)}{L} = \frac{2L}{3L} = \frac{2}{3}$$

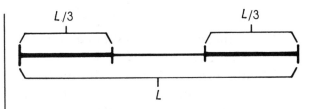

EXAMPLE 18: If a 12-inch ruler is broken in any two points at random, what is the probability that a triangle can be formed with the 3 segments?

SOLUTION: Let $x$ be the length of the left-hand segment, and let $y$ be the length of the middle segment, so that $12 - x - y$ is the length of the third segment. Then, since any side of a triangle must be less than the sum of the other two sides, the condition of this problem will be satisfied only if $x$, $y$, and $12 - x - y$ are all less than 6:

$$x < 6, y < 6, 12 - x - y < 6$$

Now, to construct a geometric diagram which expresses these algebraic requirements, we measure all possible lengths of $x$ along a horizontal 12-inch scale from $O$ to $L$, and measure corresponding lengths of $y$ and of $12 - x$ along a vertical 12-inch scale from $O$ to $M$, perpendicular to $OL$ at $O$, as in the accompanying figure. Then,

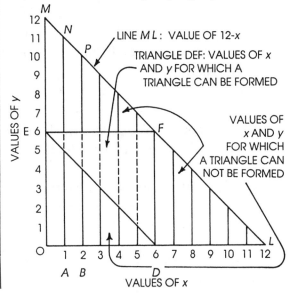

corresponding to the point $A$ for $x = 1$ inch on $OL$, we have an 11-inch vertical line $AN$ representing $12 - x = 12 - 1 = 11$, perpendicular to $OL$ at $A$; and beginning from $A$ we can find points on $AN$ corresponding to any possible values of $y$. Of the latter, however, only those between $y = 5$ and $y = 6$, dotted in the figure, satisfy the above inequalities. Likewise, corresponding to the point $B$ for $x = 2$ inches on $OL$, we have 10-inch vertical line $BP$ representing $12 - x = 12 - 2$; but on this line only those points representing values of $y$ from $y = 4$ inches to $y = 6$ inches, dotted in the figure, correspond to values of $y$ which satisfy the above equalities when $x = 2$, etc. Moreover, for all values of $x$ on $OL$ from $D$ for $x = 6$ inches, to $L$ for $x = 12$ inches, there is no corresponding value of $y$ which satisfies the above inequalities, since $x$ already violates the first inequality.

Thus we see that for each possible pair of values of $x$ and $y$ there corresponds some point in the large triangle $OLM$ with an area $= 12 \cdot \frac{12}{2} = 72$, and that for those values of $x$ and $y$ which satisfy the conditions of our problem the points all lie within the smaller triangle $DEF$ with an area $= 6 \cdot \frac{6}{2} = 18$. Hence, our required probability is:

$$p = \frac{18}{72}$$

$$= \frac{1}{4}, \text{ or one chance in four}$$

The same reasoning would apply, of couse, to a stick of any length. The only difference is that we should find it more awkward to refer repeatedly to different fractional parts of its length $L$.

---

**PRACTICALLY SPEAKING 13.3**

Mark, Steve, Howard, and Selma are playing a card game. All four have been dealt two cards from a deck of fifty-two cards. No one has been dealt a heart yet.

1. What is the probability that the next card dealt will be a heart?
2. Is it more likely, or less likely, that a heart will be dealt now, rather than when the first card was dealt?

See Appendix F for the answers.

## Exercise Set 13.3

1. Letting the heads and tails faces of three coins be designated $H_1T_1$, $H_2T_2$, $H_3T_3$, make a systematic array of the $w = n^k = 2^3 = 8$ possible combinations in which they can fall face up when tossed. Then find the corresponding values of $h$ and $f$ for the following events:

   $E_1$—all three land heads

   $E_2$—only two land heads

   $E_3$—only one lands heads

   $E_4$—none lands heads

   $E_5$—at least two land heads

2. Using your answers to Question 1, find the probabilities $p$ and $q$ of these same events happening and not happening.

3. Which of the events defined in Question 1 are mutually exclusive? Verify your answer by applying probability addition formulas.

4. For events $E_1$, $E_2$, and $E_5$, defined as in Question 1, compute $p_5$ in terms of $p_1$ and $p_2$.

5. For the events in Question 1, what are the odds

   (a) for $E_1$ to happen?

   (b) against $E_2$ to happen?

   (c) for $E_5$ to happen?

6. If a shuffled deck of cards is split at random, what is the probability that the exposed card is: (a) a "face card"—king, queen, or jack, in any suit? (b) a "black card"—spades or clubs? (c) a "black face card"? (d) either a "face card" or a "black card"?

7. What is the probability that, if five coins are tossed, all will land with the same face—heads or tails—up?

8. Each of two boxes contains ten balls which are identical except that, in each case, four are red and six are white. What is the probability that, if two balls are selected at random from each box, all four turn out to be red?

9. What is the probability in the preceding question if all four balls are drawn from the same box without any being returned?

10. Referring to the Table of Mathematical Odds for Hands Drawn in Poker, what is the probability of a player's being dealt an "opening hand"—defined as one which has a pair of jacks or any higher combination?

11. A common (Monte Carlo) type of roulette wheel has thirty-seven sectors numbered from 0 to 36, the zero being a "free house number." A player receives $36 for each dollar he bets on any given number if the wheel stops spinning with a small ball coming to rest on that number. On the sound, but here irrelevant, theory that "the house always wins," Mr. Sloe decides to place a $10 bet on the "house number" zero, rather than on 13 which he regards as "unlucky." Discuss his mathematical expectations in both cases.

12. From the solution of EXAMPLE 16 of the text, we know that the probabiity of a 4, 4, 3, 2 bridge-hand distribution is $p_1 = 0.2155$. The corresponding probabilities of 4, 3, 3, 3 and 4, 4, 4, 1 bridge-hand distributions are $p_2 = 0.1054$ and $p_3 = 0.0299$, respectively. What is the probability of drawing a hand

with a distribution of at least five cards in at least one suit?

13. One or more balls are drawn at random from a bag containing seven. What is the probability that an even number are drawn?

14. Due to a purely accidental failure of its power source, an electric clock has stopped. What is the (continuous) probability (a) that the hour hand has stopped between the 12- and 1-hour marks? (b) that both hands have stopped there?

## 13.4 Statistical Probability

In some types of problems to which we may apply the concepts of probability theory, the values of $h$ and $f$ are not derived from theoretical assumptions, but are learned statistically from experience. Hence, the latter are distinguished as cases of **statistical, empirical,** or **inductive probability,** in contrast to the nonstatistical or deducive probabilities with which we have been concerned before.

At the right, for instance, is a *Mortality Table* taken from the American Experience Table of Mortality, which insurance companies use in computing premiums for life insurance. Obviously, there is no way of knowing in advance how long any particular ten-year-old child will live, and it is reasonable to assume that those who have serious organic defects, or who grow up to pursue dangerous occupations, may very well, on the average, die sooner than others. Nevertheless, this table has been compiled, and reduced to a common denominator of 100,000, from actual statistics concerning how long people do live. By it

we see, for instance, that of $w = 100,000$ people alive at age ten, only $h = 14,474$ are still alive at age eighty. Hence, we may say, in this statistical sense, that the probability of a child of ten living to attain the age of eighty is, on the average

$$p = h/w = \frac{14,474}{100,000}$$

$$= 14.5\% \text{ approximately}$$

**MORTALITY TABLE**

| AGE | NUMBER LIVING | AGE | NUMBER LIVING |
|-----|---------------|-----|---------------|
| 10 | 100,000 | 60 | 57,917 |
| 20 | 92,637 | 70 | 38,569 |
| 30 | 85,441 | 80 | 14,474 |
| 40 | 78,106 | 90 | 847 |
| 50 | 69,804 | 100 | 0 |

Of course, the last entry in this table means only that, corresponding to each person still alive at age one hundred, there were more than 200,000 alive at age ten. Otherwise, the entry would have rounded off to 1 or more, instead of to zero, as the nearest whole number.

EXAMPLE 1: To the nearest tenth of 1%, what is the statistical probability that a child living at the age of ten (a) will still be living at age sixty? (b) will not be living at age sixty? (c) will die between the ages of sixty and seventy?

SOLUTION: (a) From the above *Mortality Table*, of $w = 100,000$ children living at age ten, only $h = 57,917$ are still living at age sixty. Hence, the probability of the latter event is:

$$p = h/w$$

$$= \frac{57,917}{100,000} = 57.9\%$$

(b) From the same figures, the number not living at age sixty is:

$$f = w - h$$
$$= 100,000 - 57,917 = 42,083$$

Therefore, the probability of such a child not living at age sixty is

$$q = f/w$$
$$= \frac{42,083}{100,000} = 42.1\%$$

Or, alternatively:

$$q = 1 - p = 1 - 57.9 = 42.1\%$$

(c) From the same *table*, the number who die between ages sixty and seventy is:

$$h = 57,917 - 38,569 = 19,348$$

Hence, the required statistical probability is:

$$p = h/w$$
$$= \frac{19,348}{100,000} = 19.3\%$$

The amounts of most insurance premiums are based on statistically determined mathematical expectations plus prorated overhead costs and reserve or profit margins. Of this, the following is a greatly simplified illustration.

EXAMPLE 2: An insurance company knows from statistical studies that 45 out of every 10,000 houses in a particular area are destroyed annually by fire. A man in this area applies for a $20,000 fire-insurance policy on his home. What must the company charge him annually for this risk in addition to its prorated other costs?

SOLUTION: The company should charge the man, for this part of its premium, the mathematical expectation which he would be purchasing. The amount is $A = \$20,000$, and the statistical probability of loss in any year is:
$$p = \frac{45}{10,000} = 0.0045$$

Therefore, the proper annual charge (for risk only) is the mathematical expectation

$$V = p \cdot A = 0.0045(\$20,000) = \$90$$

## Exercise Set 13.4

**1.** What is the statistical probability, to the nearest tenth of 1%, that of any two given persons alive at age ten will both be alive at age fifty? Use the Mortality Table in this section.

## Chapter 13 Glossary

$C(n,n) = 1$   The number of all possible combinations for the special case of $t = n$.

$C(n,t) = \dfrac{n!}{t!(n-t)!}$   The number of all possible combinations of $n$ things taken at a time.

$C(n,t) = \dfrac{p(n, t)}{t!}$   Basic relationship between $P$ and $C$.

**Combinations**   Groupings of things without paying attention to order.

**Dependent Events**   Two separate events that do affect the happening of one another.

**Factorial**   A product symbolized by $n!$.

**Independent Events**   Two separate events that do not affect the happening of one another.

$M = c(k,1) + C(k,2) + \ldots C(k,k) = 2_k - 1$   The number of all possible combinations of $k$ things taken 1, or 2, or . . . $k$ at a time.

**Mathematical Expectation** The product of the probability, P, that a particular event will take place, and the amount, $A$, one will receive if it does happen.

**Mutually Exclusive** Two events are mutually exclusive if the happening of either excludes the possible happening of the other on the same trial.

$N = C(n_1,t_1)C(n_2,t_2) \ldots C(n_k,t_k)$ The number of all possible intergroup combinations of $n_1$ things taken $t_1$ at a time, and so on, up to $n_k$ things taken $t_k$ at a time.

$N = n_1n_2n_3 \ldots n_k$ The number of all possible intergroup combinations for the special case of $t_1 = t_2 = \ldots t_k = 1$.

$N = n^k$ The number of all possible intergroup combinations for the special case of $n_1 = n_2 = \ldots n_k = n$.

$P_c(n,n) = (n-1)!$ The number $P_c$ of all possible circular permutations of $n$ things arranged in a circle or closed chain.

$$P_a = \frac{n!}{n_1!n_2! \ldots}$$ The number $P_a$ of all possible indistinguishable permutations of $n$ things when $n_1$ are all alike, $n_2$ are all alike, and so on.

$P(n,n) = n!$ The number $P$ of all possible permutations in the special case when $t = n$.

$$P(n,t) = \frac{n!}{(n-t)!}$$ The number $P$ of all possible permutations of any $n$ things taken $t$ at a time.

$P(n,t) = C(n,t)P(t,t)$ Basic relationship between $P$ and $C$.

**Permutations** Arrangements of things in a specific order.

## Chapter 13 Test

For each problem, five answers are given. Only one answer is correct. After you solve each problem, check the answer that agrees with your solution.

1. How many ways are there to arrange six books on a shelf?

   A) $P(6,6)$     D) $C(1,1)$
   B) $C(6,6)$     E) $P(1,6)$
   C) $P(6,1)$

2. In how many ways can three pieces of furniture be arranged along a wall?

   A) $P(1,1)$     D) $C(3,1)$
   B) $P(2,1)$     E) $P(3,3)$
   C) $C(3,3)$

3. In how many ways can eight cards be dealt from a deck of fifty-two cards?

   A) $C(52,8)$    D) $C(8,52)$
   B) $C(13,8)$    E) $C(8,1)$
   C) $P(52,8)$

4. In how many ways can fifteen cards be dealt from a deck of fifty-two cards?

   A) $P(52,15)$   D) $P(15,1)$
   B) $C(15,1)$    E) $(52-1)!$
   C) $C(52,15)$

5. In how many ways can sixteen people sit at a circular table?

   A) 16!          D) $C(16,16)$
   B) 15!          E) $C(16,1)$
   C) $P(16,12)$

6. In how many ways can nine people sit on one side of a long table?

   A) $C(9,9)$     D) $C(1,9)$
   B) $P(9,9)$     E) $P(9,1)$
   C) $C(9,9)$

7. In how many ways can a team of nine softball players be chosen from a group of seventeen people?

   A) $P(9,9)$     D) $P(17,9)$
   B) $C(17,17)$   E) $C(17,9)$
   C) $C(9,17)$

8. In how many ways can a baseball bat be chosen from a pile of four different baseball bats?

   **A)** $P(4,2)$     **D)** $C(4,1)$
   **B)** $P(4,1)$     **E)** $C(1,1)$
   **C)** $P(1,1)$

9. In how many ways can Charles line up six cans of soda on a cooler? Two of the cans are colas, and the other four are ginger ales.

   **A)** $\dfrac{9!}{2!4!}$     **D)** $\dfrac{6!2!}{4!}$
   **B)** $\dfrac{6!}{2!4!}$     **E)** $\dfrac{2!}{6!4!}$
   **C)** $\dfrac{4!}{6!}$

10. In how many ways can six soldiers stand in line at attention?

    **A)** $P(1,1)$     **D)** $C(6,6)$
    **B)** $P(6,1)$     **E)** $C(6,1)$
    **C)** $P(6,6)$

11. In how many ways can five people sit at a round table in a restaurant?

    **A)** $(5 - 1)!$     **D)** $P(5,5)$
    **B)** $C(5,1)$     **E)** $5!$
    **C)** $(5 + 1)!$

12. In how many ways can Jim sit at a round table with six of his friends if he always sits next to his best friend Mark?

    **A)** $(5 - 1)!$     **D)** $C(6,1)$
    **B)** $(6 - 1)!$     **E)** $P(6,1)$
    **C)** $(7 - 1)!$

13. In how many ways can Albert place three different books on a shelf?

    **A)** $C(3,3)$     **D)** $P(3,1)$
    **B)** $P(3,3)$     **E)** $C(3,1)$
    **C)** $(3 - 1)!$

14. In how many ways can Jackie stack five different party invitations?

    **A)** $(5 - 1)!$     **D)** $P(5,5)$
    **B)** $P(5,1)$     **E)** $C(5,1)$
    **C)** $C(5,5)$

15. What is the probability of having three coins in a row land heads up when tossed?

    **A)** $\frac{1}{32}$     **D)** $\frac{1}{16}$
    **B)** $\frac{1}{8}$     **E)** $\frac{1}{64}$
    **C)** $\frac{1}{4}$

16. What is the probability of being dealt a heart when dealt one of fifty-two cards?

    **A)** $\frac{4}{13}$     **D)** $\frac{1}{13}$
    **B)** $\frac{1}{52}$     **E)** $\frac{13}{26}$
    **C)** $\frac{13}{52}$

17. What is the probability of having a die land with a five showing?

    **A)** $\frac{5}{6}$     **D)** $\frac{1}{6}$
    **B)** $\frac{4}{6}$     **E)** $\frac{5}{12}$
    **C)** $\frac{2}{6}$

18. If two dice are tossed, what is the probability of the sum showing being four?

    **A)** $\frac{3}{36}$     **D)** $\frac{7}{36}$
    **B)** $\frac{2}{4}$     **E)** $\frac{4}{6}$
    **C)** $\frac{4}{36}$

19. Wht s the probability of being dealt two aces when dealt two cards in succession from a fifty-two card deck?

    **A)** $\frac{2}{26}$     **D)** $\frac{1}{221}$
    **B)** $\frac{2}{52}$     **E)** $\frac{1}{13}$
    **C)** $\frac{1}{36}$

20. What is the probability of getting a heart and then a club when dealt two cards in succession from a fifty-two card deck?

    **A)** $\frac{2}{52}$     **D)** $\frac{2}{26}$
    **B)** $\frac{13}{204}$     **E)** $\frac{26}{52}$
    **C)** $\frac{13}{52}$

# Trigonometry

## 14.1 Trigonometric Functions

**Trigonometry** is the branch of mathematics that deals with the measurement of triangles. (The word *trigonometry* comes from the Greek and means *to measure a triangle.*) Trigonometry enables us to find the unknown parts of triangles by arithmetical processes. For this reason it is constantly used in surveying, mechanics, navigation, engineering, physics, and astronomy.

In geometry we learned that there are many shapes of triangles. For our purpose we can start with the simple case of a right triangle. Starting from this, we will eventually be able to work with all types of triangles because any triangle can be broken down into two right triangles.

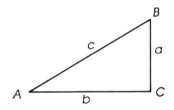

In the right triangle *BAC* we know from geometry that

**a)** $\angle A + \angle B = 90$

**b)** $c^2 = a^2 + b^2$

From equation *a* we can find one of the acute angles if the other is given, and from equation *b* we can determine the length of any side if the other two are given. But as yet we do not have a method for finding $\angle A$ if given the two sides *a* and *b*, even though by geometry we could construct the triangle with this information. And this is where trigonometry makes its contribution. It gives us a method for calculating the angles if we know the sides or for calculating the sides if we know the angles.

## Trigonometric Functions of an Angle

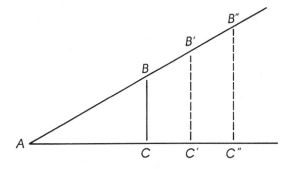

If we take the triangle in the previous figure and extend lines $AB$ and $AC$, and then drop perpendiculars from points $B'$ and $B''$ to $AC$, we form three similar triangles:

$$CAB, \ C' AB', \text{ and } C'' AB''$$

When two triangles are similar, the ratio of any two sides of one triangle equals the ratio of corresponding sides of the second triangle. Thus, in the three triangles of the figure

$$\frac{BC}{AC} = \frac{B'C'}{AC'} = \frac{B''C''}{AC''}, \text{ or}$$

$$\frac{BC}{AB} = \frac{B'C'}{AB'} = \frac{B''C''}{AB''}$$

Similar equalities hold for the ratios between the other sides of the triangles.

These equalities between the ratios of the corresponding sides of similar triangles illustrate the fact that *no matter how the size of a right triangle may vary, the val-*

*ues of the ratios of the sides remain the same so long as the acute angles are unchanged.* In other words each of the above ratios is a **function** of $\angle A$.

From algebra and geometry we learn that a variable quantity which depends upon another quantity for its value is called a **function** of the latter value.

Therefore in the above figure the value of the ratio $\dfrac{BC}{AC}$ is a function of the magnitude of $\angle A$; and as long as the magnitude of $\angle A$ remains the same, the value of the ratio $\dfrac{BC}{AC}$ will be the same.

## Description of the Tangent Function

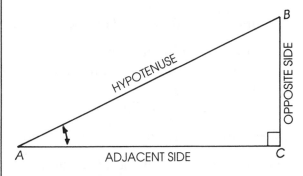

The constant ratio or function $\dfrac{BC}{AC}$ is termed the **tangent** of $\angle A$. It will be noted that this function represents the ratio of the side *opposite* $\angle A$ divided by the side next to $\angle A$, called the *adjacent* side—that is, the side next to it other than the hypotenuse. Accordingly:

$$\text{tangent } \angle A = \frac{\text{opposite side}}{\text{adjacent side}}$$

$$\text{or} \quad \tan A = \frac{\text{opp}}{\text{adj}}$$

## Making a Table of Trigonometric Functions

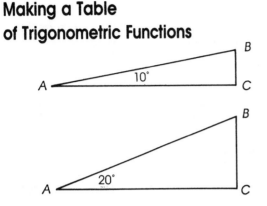

If we construct $\angle A$ equal to 10° and measure $BC$ and $AC$ and then compute the value of $\dfrac{BC}{AC}$, we will find it to be .176. Then if we construct $\angle A$ to equal 20°, we will find $\dfrac{BC}{AC}$ equal to .364. For $\angle A$ at 30° we will find $\dfrac{BC}{AC}$ equal to .577. This means that thereafter we will know that the *tangent* of any angle of 10° in a right triangle is equal to .176, and the tangent of any angle of 20° is equal to .364. Thus, by computing the values of the ratios of $\dfrac{BC}{AC}$ for all angles from 1° to 90° we would obtain a complete table of tangent values. A sample of such a table is shown below.

**SAMPLE TABLE OF TRIGONOMETRIC FUNCTIONS**

| ANGLE | SINE | COSINE | TANGENT |
|---|---|---|---|
| 68° | .9272 | .3746 | 2.4751 |
| 69° | .9336 | .3584 | 2.6051 |
| 70° | .9397 | .3420 | 2.7475 |
| 71° | .9455 | .3256 | 2.9042 |
| 72° | .9511 | .3090 | 3.0777 |

This sample table gives the tangents of angles to four decimal places. For instance, in the table above, to find the value of the tangent of an angle of 69° we first look in the column headed *Angle* and find 69°. Then on the same horizontal line in the column headed *Tangent* we find the value 2.6051. This means that tan 69° = 2.6051.

The following example will show us how we can solve problems in trigonometry by the use of the table of tangents.

EXAMPLE 1: An airplane is sighted by two observers. One observer at *A* indicates it to be directly overhead. The other observer at *B*, 3,000 feet due west of *A*, measures its angle of elevation (*see below*) at 70°. What is the altitude of the airplane?

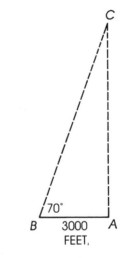

SOLUTION:

$$\tan \angle B = \frac{\text{(opp side)}}{\text{(adj side)}} = \frac{CA}{BA}$$

Since $\qquad \angle B = 70°$

$$\tan \angle B = 2.7475$$

$$2.7475 = \frac{CA}{3,000} \quad \text{Substitute.}$$

$$CA = 3,000 \times 2.7475$$

Transpose.

$$= 8,242.5 \text{ feet}$$

Altitude of the airplane is 8242.5 feet.

## Practical Observation of Angles

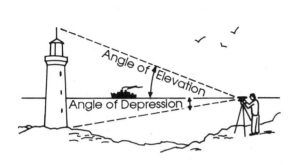

The *angle of elevation or depression* of an object is the angle made between a line from the eye to the object and a horizontal line in the same vertical plane. If the object is above the horizontal line it makes an *angle of elevation;* if below the horizontal line it makes an *angle of depression.*

## The Six Trigonometric Functions

As has been previously pointed out, ratios other than those involved in the *tangent function* exist between the sides of the triangle, and have, like the tangent, an equality of value for a given magnitude of angle, irrespective of the size of the triangle. It is to be expected, therefore, that problems involving the solution of right triangles can be solved by other known trigonometric ratios or functions of the selfsame angle. There are six important ratios or functions for any acute angle of a right triangle. The description and definition of these functions follows.

The sides and angles of triangle *CAB* in the following diagram have been marked in the manner traditionally employed in trigonometry. It is the custom to have the angles represented by capital letters and the sides indicated by the small letter corresponding to the angle opposite the side.

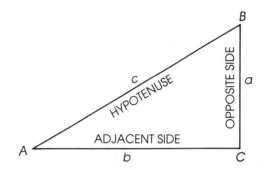

Thus the right angle is designated by *C* while the hypotenuse, which is opposite to it, is designated by *c*. Similarly, side *a* is opposite ∠*A*, and side *b* is opposite ∠*B*. Thus we have these six ratios:

$\dfrac{a}{c}$ is the **sine** of ∠*A* (written sin *A*)

$\dfrac{b}{c}$ is the **cosine** of ∠*A* (written cos *A*)

$\dfrac{a}{b}$ is the **tangent** of ∠*A* (written tan *A*)

$\dfrac{b}{a}$ is the **cotangent** of ∠*A* (written cot *A*)

$\dfrac{c}{b}$ is the **secant** of ∠*A* (written sec *A*)

$\dfrac{c}{a}$ is the **cosecant** of ∠*A* (written csc *A*)

Using self-explanatory abbreviations, we have by definition:

$$\sin A = \frac{\text{opp}}{\text{hyp}} = \frac{a}{c} \qquad \cos A = \frac{\text{adj}}{\text{hyp}} = \frac{b}{c}$$

$$\tan A = \frac{\text{opp}}{\text{adj}} = \frac{a}{b} \qquad \cot A = \frac{\text{adj}}{\text{opp}} = \frac{b}{a}$$

$$\sec A = \frac{\text{hyp}}{\text{adj}} = \frac{c}{b} \qquad \csc A = \frac{\text{hyp}}{\text{opp}} = \frac{c}{a}$$

These definitions of the trigonometric functions should be memorized.

## Exercise Set 14.1

1. In the preceding figure, $\tan B = \dfrac{b}{a}$.

   Write the other five functions of $\angle B$.
2. Which is greater, $\sin A$ or $\tan A$?
3. Which is greater, $\cos A$ or $\cot A$?
4. Which is greater, $\sec A$ or $\tan A$?
5. Which is greater, $\csc A$ or $\cot A$?
6. Sin $A = \frac{3}{5}$. What is the value of cos $A$?

   *Hint:* Use the right triangle formula $c^2 = a^2 + b^2$ to find side $b$.
7. Tan $A = \frac{3}{4}$. What is the value of sin $A$?
8. Sin $A = \frac{8}{17}$. Find cos $A$.
9. Cot $A = \frac{15}{8}$. Find sec $A$.
10. Find the value of the other five functions of $A$ if $\sin A = \frac{5}{13}$.

## 14.2 Relations Between Functions of Complementary Angles

If we observe the relations between the functions of the two acute angles of the same right triangle, we will note that every function of each of the two acute angles is equal to a different function of the other acute angle. These correspondences of value are demonstrated in the following.

$$\sin A = \frac{a}{c} \quad \text{and} \quad \cos B = \frac{a}{c}$$

$$\cos A = \frac{b}{c} \quad \text{and} \quad \sin B = \frac{b}{c}$$

$$\tan A = \frac{a}{b} \quad \text{and} \quad \cot B = \frac{a}{b}, \text{etc.}$$

Thus we have:

$$\sin A = \cos B, \qquad \cot A = \tan B$$

$$\cos A = \sin B, \qquad \sec A = \csc B$$

$$\tan A = \cot B, \qquad \csc A = \sec B$$

From these equalities it will be evident that any function of an acute angle of a right triangle equals the cofunction of the complement of that angle.*

For example, $\tan 40° = \cot 50°$; $\sin 70° = \cos 20°$; $\csc 41° 20' = \sec 48° 40'$.

Since angles $A$ and $B$ are complementary, another way of writing these equations is as follows:

$$\sin (90° - A) = \cos A$$

$$\cos (90° - A) = \sin A$$

$$\tan (90° - A) = \cot A$$

$$\cot (90° - A) = \tan A$$

$$\sec (90° - A) = \csc A$$

$$\csc (90° - A) = \sec A$$

## Exercise Set 14.2

Find the equivalent cofunctions of the following trigonometric functions for questions 1 through 6.

1. $\sin 26° =$
2. $\tan 43° =$
3. $\cos 24° 28' =$
4. $\cot 88° 50' =$
5. $\sec 6° 10' =$
6. $\csc 77\frac{1}{2}° =$
7. How many degrees must $\angle A$ be if $90° - A = 5A$?
8. What is the value of $\angle A$ if $\tan A = \cot A$?
9. Find $A$ if $90° - A = A$.
10. Find $A$ if $\cos A = \sin 2A$.

* The name *cosine* means *complement's sine*. It is a contraction from the Latin *complementi sinus*. The words *cotangent* and *cosecant* were derived in the same manner.

# 14.3 Using a Trigonometric Function Table

**TABLE OF NATURAL TRIGONOMETRIC FUNCTIONS**

| ANGLE | SIN | COS | TAN | COT | SEC | CSC | |
|---|---|---|---|---|---|---|---|
| 0° | .0000 | 1.0000 | .0000 | ∞ | 1.0000 | ∞ | 90° |
| 1 | .0175 | .9998 | .0175 | 57.2900 | 1.0002 | 57.2987 | 89 |
| 2 | .0349 | .9994 | .0349 | 28.6363 | 1.0006 | 28.6537 | 88 |
| 3 | .0523 | .9986 | .0524 | 19.0811 | 1.0014 | 19.1073 | 87 |
| 4 | .0698 | .9976 | .0699 | 14.3007 | 1.0024 | 14.3356 | 86 |
| 5° | .0872 | .9962 | .0875 | 11.4301 | 1.0038 | 11.4737 | 85° |
| 6 | .1045 | .9945 | .1051 | 9.5144 | 1.0055 | 9.5668 | 84 |
| 7 | .1219 | .9925 | .1228 | 8.1443 | 1.0075 | 8.2055 | 83 |
| 8 | .1392 | .9903 | .1405 | 7.1154 | 1.0098 | 7.1853 | 82 |
| 9 | .1564 | .9877 | .1584 | 6.3138 | 1.0125 | 6.3925 | 81 |
| 10° | .1736 | .9848 | .1763 | 5.6713 | 1.0154 | 5.7588 | 80° |
| 11 | .1908 | .9816 | .1944 | 5.1446 | 1.0187 | 5.2408 | 79 |
| 12 | .2079 | .9781 | .2126 | 4.7046 | 1.0223 | 4.8097 | 78 |
| 13 | .2250 | .9744 | .2309 | 4.3315 | 1.0263 | 4.4454 | 77 |
| 14 | .2419 | .9703 | .2493 | 4.0108 | 1.0306 | 4.1336 | 76 |
| 15° | .2588 | .9659 | .2679 | 3.7321 | 1.0353 | 3.8637 | 75° |
| 16 | .2756 | .9613 | .2867 | 3.4874 | 1.0403 | 3.6280 | 74 |
| 17 | .2924 | .9563 | .3057 | 3.2709 | 1.0457 | 3.4203 | 73 |
| 18 | .3090 | .9511 | .3249 | 3.0777 | 1.0515 | 3.2361 | 72 |
| 19 | .3256 | .9455 | .3443 | 2.9042 | 1.0576 | 3.0716 | 71 |
| 20° | .3420 | .9397 | .3640 | 2.7475 | 1.0642 | 2.9238 | 70° |
| 21 | .3584 | .9336 | .3839 | 2.6051 | 1.0711 | 2.7904 | 69 |
| 22 | .3746 | .9272 | .4040 | 2.4751 | 1.0785 | 2.6695 | 68 |
| 23 | .3907 | .9205 | .4245 | 2.3559 | 1.0864 | 2.5593 | 67 |
| 24 | .4067 | .9135 | .4452 | 2.2460 | 1.0946 | 2.4586 | 66 |
| 25° | .4226 | .9063 | .4663 | 2.1445 | 1.1034 | 2.3662 | 65° |
| 26 | .4384 | .8988 | .4877 | 2.0503 | 1.1126 | 2.2812 | 64 |
| 27 | .4540 | .8910 | .5095 | 1.9626 | 1.1223 | 2.2027 | 63 |
| 28 | .4695 | .8829 | .5317 | 1.8807 | 1.1326 | 2.1301 | 62 |
| 29 | .4848 | .8746 | .5543 | 1.8040 | 1.1434 | 2.0627 | 61 |
| 30° | .5000 | .8660 | .5774 | 1.7321 | 1.1547 | 2.0000 | 60° |
| 31 | .5150 | .8572 | .6009 | 1.6643 | 1.1666 | 1.9416 | 59 |
| 32 | .5299 | .8480 | .6249 | 1.6003 | 1.1792 | 1.8871 | 58 |
| 33 | .5446 | .8387 | .6494 | 1.5399 | 1.1924 | 1.8361 | 57 |
| 34 | .5592 | .8290 | .6745 | 1.4826 | 1.2062 | 1.7883 | 56 |
| 35° | .5736 | .8192 | .7002 | 1.4281 | 1.2208 | 1.7434 | 55° |
| 36 | .5878 | .8090 | .7265 | 1.3764 | 1.2361 | 1.7013 | 54 |
| 37 | .6018 | .7986 | .7536 | 1.3270 | 1.2521 | 1.6616 | 53 |
| 38 | .6157 | .7880 | .7813 | 1.2799 | 1.2690 | 1.6243 | 52 |
| 39 | .6293 | .7771 | .8098 | 1.2349 | 1.2868 | 1.5890 | 51 |
| 40° | .6428 | .7660 | .8391 | 1.1918 | 1.3054 | 1.5557 | 50° |
| 41 | .6561 | .7547 | .8693 | 1.1504 | 1.3250 | 1.5243 | 49 |
| 42 | .6691 | .7431 | .9004 | 1.1106 | 1.3456 | 1.4945 | 48 |
| 43 | .6820 | .7314 | .9325 | 1.0724 | 1.3673 | 1.4663 | 47 |
| 44 | .6947 | .7193 | .9657 | 1.0355 | 1.3902 | 1.4396 | 46 |
| 45° | .7071 | .7071 | 1.0000 | 1.0000 | 1.4142 | 1.4142 | 45° |
| | COS | SIN | COT | TAN | CSC | SEC | ANGLE |

From the table on the previous page, it becomes apparent that we can easily compute the functions of any angle greater than 45° if we know the functions of all angles between 0° to 45°. Therefore in a table of trigonometric functions, it is only necessary to have a direct table of functions for angles from 0° to 45°, since the function of any angle above 45° is equal to the cofunction of its complement.

To find the functions of angles from 0° to 45° read the table from the top down, using the values of angles at the left and the headings at the top of the table. To find the functions of angles from 45° to 90° read from the bottom up, using the values of angles at the right and the function designations at the bottom of the table.

If we know the value of the function of an angle and wish to find the angle, look in the body of the table in the proper column and then read the magnitude of the angle in the corresponding row of one or the other of the angle columns.

EXAMPLE 1: If the sine of an angle is .5000, find the angle.

SOLUTION: Look in the *Sin* column, locate .5000, and read the angle value (30°) from the left *Angle* column. If this value had been given to us as a cosine, we would have noted that it does not appear in the column headed *Cos* at the top but does appear in the column that has *Cos* at the bottom. Therefore, we would then use the *Angle* column at the right and find .5000 to be the cosine of 60°.

If we inspect the ratios of the six functions of $\angle A$, we will note that they are not independent of each other. In fact, if we line them up as follows:

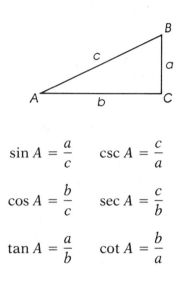

$$\sin A = \frac{a}{c} \qquad \csc A = \frac{c}{a}$$

$$\cos A = \frac{b}{c} \qquad \sec A = \frac{c}{b}$$

$$\tan A = \frac{a}{b} \qquad \cot A = \frac{b}{a}$$

it becomes obvious that the sine is the reciprocal of the cosecant, the cosine is the reciprocal of the secant, and the tangent is the reciprocal of the cotangent. Accordingly:

$$\sin A = \frac{1}{\csc A} \qquad \cos A = \frac{1}{\sec A}$$

$$\tan A = \frac{1}{\cot A} \qquad \csc A = \frac{1}{\sin A}$$

$$\sec A = \frac{1}{\cos A} \qquad \cot A = \frac{1}{\tan A}$$

Therefore:

$\sin A \times \csc A = 1$, $\cos A \times \sec A = 1$

$\tan A \times \cot A = 1$

In accordance with the usual algebraic method of notation (by which *ab* is equivalent to $a \times b$) these relationships are usually written:

$$\sin A \csc A = 1, \cos A \sec A = 1$$

$$\tan A \cot A = 1$$

To illustrate such a relation, find, for example, in the table of functions the tangent and the cotangent of 30°.

$$\tan 30° = .5774, \cot 30° = 1.7321$$

$$\tan 30° \cot 30° = .5774 \times 1.7321$$
$$= 1.00011454$$

## Interrelations Among the Functions

Since $\tan A = \dfrac{a}{b}$, $\sin A = \dfrac{a}{c}$, and $\cos A = \dfrac{b}{c}$, it follows that

$$\tan A = \frac{\sin A}{\cos A}, \text{ and } \sin A = \tan A \cos A$$

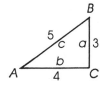

From the interrelations of sine, cosine, and tangent it follows that if we know two of these values, we can always find the third.

From the Pythagorean theorem of the right triangle we know that $a^2 + b^2 = c^2$. If we divide both sides of this equation by $c^2$, we get

$$\frac{a^2}{c^2} + \frac{b^2}{c^2} = 1$$

Since $\dfrac{a}{c} = \sin A$ and $\dfrac{b}{c} = \cos A$, it follows that

$$(1)\ \sin^2 A + \cos^2 A = 1†$$

† $(\sin A)^2$ is customarily written as $\sin^2 A$, and likewise for the other functions.

Therefore:

$$(2)\ \sin A = \sqrt{1 - \cos^2 A} \quad \text{and}$$

$$(3)\ \cos A = \sqrt{1 - \sin^2 A}$$

## Making Practical Use of the Functions

With the information on trigonometry outlined in the previous pages we will be able to solve many triangles if we know three parts, one of which is a side. And in the case of the right triangle, since the right angle is a part of it, we need only to know two other parts, one of which must be a side.

As will be brought out in the practice exercises that follow, these trigonometric methods of solving triangles are used daily in handling problems that arise in military operations, engineering, navigation, shopwork, physics, surveying, etc.

We should adopt a planned method of procedure in solving problems. One such method is as follows.

1. After reading the problem, draw a figure to a convenient scale, and in it show those lines and angles which are given and those which are to be found.

2. Write down all the formulas that apply to the particular problem.

3. Substitute the given data in the proper formulas, and solve for the unknowns.

4. Check your results.

$$\sin A = \frac{\text{opp}}{\text{hyp}} = \frac{a}{c}$$

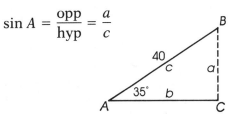

EXAMPLE 2: In the accompanying figure $c = 40$ and $\angle A = 35°$. Find $a$.

SOLUTION: $\dfrac{a}{c} = \sin A$, $a = c \sin A$

$\sin 35° = .5736$, $c = 40$

$c \sin A = 40 \times .5736 = 22.944$

$a = 22.944$

Check: $\dfrac{a}{c} = \sin A$

$\dfrac{22.944}{40} = .5736$, which is $\sin 35°$

EXAMPLE 3: Given $c = 48$ and $\angle B = 22°$, find $a$ by means of the sine formula.

SOLUTION: $\dfrac{a}{c} = \sin A$, $a = c \sin A$

$\angle A = 90° - \angle B$

$\angle A = 90° - 22° = 68°$

$\sin 68° = .9272$, $c = 48$

$c \sin A = 48 \times .9272 = 44.5056$

$a = 44.50+$

Check: $\dfrac{a}{c} = \sin A$

$\dfrac{44.5056}{48} = .9272$, which is $\sin 68°$

$\cos A = \dfrac{\text{adj}}{\text{hyp}} = \dfrac{b}{c}$

EXAMPLE 4: In the accompanying figure, $c = 36$ and $\angle A = 40°$. Find $b$.

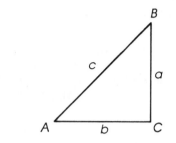

SOLUTION: $\dfrac{b}{c} = \cos A$, $b = c \cos A$

$\cos 40° = .7660$

$c = 36$

$c \cos A = 36 \times .7660 = 27.576$

$b = 27.58$

Check: $\dfrac{b}{c} = \cos A$, $\dfrac{27.576}{36} = .7660$ or cos $40°$

EXAMPLE 5: Given $b = 26$ and $\angle A = 22°$; find $c$.

SOLUTION: $\dfrac{b}{c} = \cos A$, $c = \dfrac{b}{\cos A}$

$b = 26$

$\cos 22° = .9272$

$\dfrac{b}{\cos A} = 26 \div .9272 = 28.04$

$c = 28.04$

Check: $\dfrac{b}{c} = \cos A$

$\dfrac{26}{28.04} = .9272$, which is $\cos 22°$.

$\tan A = \dfrac{\text{opp}}{\text{adj}} = \dfrac{a}{b}$

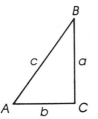

EXAMPLE 6: In the accompanying figure $a = 40$ and $b = 27$. Find $\angle A$.

SOLUTION: $\dfrac{a}{b} = \tan A$, $a = 40$, $b = 27$

$$\frac{40}{27} = 1.4815$$

$$\tan A = 1.4815, \angle A = 55° \, 59'$$

Check: $a = b \tan A$; $27 \times 1.4815 = 40$ which is $a$.

EXAMPLE 7: Given angle $A = 28°$ and $a = 29$. Find $b$.

SOLUTION: $\dfrac{a}{b} = \tan A$, $b = \dfrac{a}{\tan A}$

$a = 29$, $\tan 28° = .5317$, $\dfrac{29}{.5317} = 54.54$

$b = 54.54$

Check: $\dfrac{a}{b} = \tan A$

$$\frac{29}{54.54} = .5317, \text{ which is } \tan A.$$

## Exercise Set 14.3

From the table of trigonometric functions find the values required in examples 1 to 15:

1. sin 8°    6. cos 25°    11. cos 62°
2. sin 42°    7. csc 14°    12. tan 56°
3. tan 40°    8. sin 78°    13. sin 58°
4. cot 63°    9. cot 69°    14. cos 45°
5. sec 22°    10. sec 81°    15. sin 30°
16. Find the angle whose sine is .2588.
17. Find the angle whose tangent is .7002.
18. Find the angle whose cosine is .5000.
19. Find the angle whose secant is 2.9238.
20. Find the angle whose cotangent is 5.6713.

The problems in this exercise should be solved by using the sine function. Answers need be accurate only to the first decimal place.

21. Given $c = 100$, $\angle A = 33°$, find $a$.
22. Given $c = 10$, $\angle A = 20°$, find $a$.
23. Given $a = 71$, $c = 78$, find $\angle A$.
24. Given $a = 14$, $\angle A = 28°$, find $c$.
25. Given $c = 50$, $a = 36$, find $\angle A$.

Use the cosine function in solving the problems in this exercise.

26. Given $c = 400$, $b = 240$, find $\angle A$.
27. Given $c = 41$, $\angle A = 39°$, find $b$.
28. Given $c = 67.7$, $\angle A = 23° \, 30'$, find $b$.
29. Given $c = 187$, $b = 93\frac{1}{2}$, find $\angle A$.
30. Given $b = 40$, $\angle A = 18°$, find $c$.

Use the tangent function in solving these problems.

31. Given $a = 18$, $b = 24$, find $\angle A$.
32. Given $b = 64$, $\angle A = 45°$, find $a$.
33. Given $b = 62$, $\angle A = 36°$, find $a$.
34. Given $\angle A = 70°$, $a = 50$, find $b$.
35. Given $\angle A = 19° \, 36'$, $b = 42$, find $a$.

## 14.4 Functions of 45°, 30°, and 60° Angles

For some rather common angles the exact values of their functions can be easily found by the application of elementary principles of geometry.

### Functions of a 45° Angle

In the isosceles right triangle *ACB*, if $\angle A = 45°$, then $\angle B = 45°$, and therefore side $a$ = side $b$. Now if we let side $a$ equal 1 or unity, then from the right triangle formula of

$$a^2 + b^2 = c^2$$

we get

$$c = \sqrt{1 + 1} \text{ or } \sqrt{2}$$

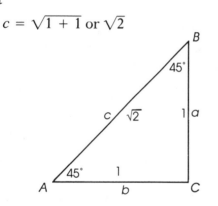

(taking the square root of both sides of the equation). Now since any trigonometric function of an acute angle is equal to the corresponding cofunction of its complement, therefore:

$$\sin 45° = \frac{1}{\sqrt{2}} \text{ or } \frac{1}{2}\sqrt{2} = \cos 45°$$

$$\tan 45° = \frac{1}{1} \text{ or } 1 = \cot 45°$$

$$\sec 45° = \frac{\sqrt{2}}{1} \text{ or } \sqrt{2} = \csc 45°$$

### Functions of 30° and 60° Angles

In the equilateral triangle *ABD* the three sides are equal and the three angles each equal 60°. If we drop a perpendicular from *B* to *AD*, it bisects $\angle B$ and the base *AD* at *C*.

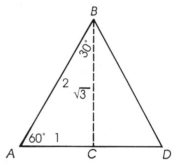

If we let the length of each of the sides equal 2 units, then $AC = CD = 1$; and in the right triangle *ACB*

$$\angle B = 30°, \angle C = 90°, \angle A = 60°$$
$$AC = 1, AB = 2$$

Then, since $(AB)^2 = (AC)^2 + (BC)^2$, it follows that $(BC)^2 = 3$ and $BC = \sqrt{3}$.

Thus in the right triangle *ACB*

$$\sin 30° = \frac{1}{2} \qquad = \cos 60°$$

$$\tan 30° = \frac{1}{\sqrt{3}} \text{ or } \frac{1}{3}\sqrt{3} = \cot 60°$$

$$\sec 30° = \frac{2}{\sqrt{3}} \text{ or } \frac{2}{3}\sqrt{3} = \csc 60°$$

$$\cos 30° = \frac{\sqrt{3}}{2} \qquad = \sin 60°$$

$$\cot 30° = \frac{\sqrt{3}}{1} \text{ or } \sqrt{3} \quad = \tan 60°$$

$$\csc 30° = \frac{2}{1} \text{ or } 2 \qquad = \sec 60°$$

It is an advantage to know the values of the 30°, 45°, and 60° angles by heart. To help yourself memorize them, fill in the outline of the table below with the proper values of the functions.

| FUNCTION | 30° | 60° | 45° |
|----------|-----|-----|-----|
| Sine | | | |
| Cosine | | | |
| Tangent | | | |
| Cotangent | | | |
| Secant | | | |
| Cosecant | | | |

## Interpolation

Interpolation is used in trigonometry in connection with the table of functions. For example, if given the function of an angle that is measured in degrees and minutes, such as sin 30° 40', its exact value could not be found directly from the table but would have to be computed by the method of interpolation. Again, if given the value of a trigonometric function such as tan $A$ = .7400, which does not appear in the body of the table, it means that the corresponding angle is expressed in units more exact than the nearest degree and must be found by interpolation. The following examples will illustrate the method of performing interpolations with reference to the table of trigonometric functions.

EXAMPLE 1: Find sin 30° 40'.

SOLUTION: Sin 30° 40' is between sin 30° and sin 31°.

Since there are 60' in 1°, 40' = $\frac{2}{3}$ of 1°

From the table      sin 30° = .5000

sin 31° = .5150

Difference = .0150

sin 30° = .5000

$\frac{2}{3}$ of .0150 = .0100

sin 30° 40' = .5100

*Note:* In this case we added the proportional part of the difference (.0100) to the value of sin 30° because the sine of an angle increases as the angle increases.

EXAMPLE 2: Find cos 59° 48'.

SOLUTION: Cos 59° 48' is between cos 59° and cos 60°. 48' is $\frac{4}{5}$ of 1°.

From the table   cos 59° = .5150

cos 60° = .5000

Difference = .0150

cos 59° = .5150

$\frac{4}{5}$ of .0150 = .0120

cos 59° 48' = .5030

*Note:* In this case we subtracted the proportional part of the difference (.0120) from the value of cos 59° because the cosine of an angle decreases as the angle increases.

EXAMPLE 3: Find $\angle A$ if tan $A = .7400$.

SOLUTION: From the table, in the tan column, we see that .7400 is between tan 36° and tan 37°.

$$\text{tan } 37° = .7536$$
$$\text{tan } 36° = \underline{.7265}$$
$$\text{Difference} = .0271$$

$$\text{tan } A = .7400$$
$$\text{tan } 36° = \underline{.7265}$$
$$\text{Difference} = .0135$$

The proportional difference between tan $A$ and tan 36° is .0135. The difference between tan 36° and tan 37° is .0271.

$$\frac{.0135}{.0271} \text{ of } 1° \text{ or } 60' \text{ equals } \tfrac{1}{2}° \text{ or } 30'$$

Therefore tan $A = 36° + 30' = 36° 30'$

Further familiarity with the table of functions will indicate the following about variations of the trigonometric functions.

As an angle increases from 0° to 90°, its:

| | |
|---|---|
| sine | increases from 0 to 1 |
| cosine | decreases from 1 to 0 |
| tangent | increases from 0 to ∞ |
| cotangent | decreases from ∞ to 0 |
| secant | increases from 1 to ∞ |
| cosecant | decreases from ∞ to 1 |

Also note that:

sines and cosines are never > 1

secants and cosecants are never < 1

tangents and cotangents may have any value from 0 to ∞.

## Exercise Set 14.4

Find by interpolation the values of the functions in problems 1 to 5.

1. sin 15° 30'
2. cos 25° 40'
3. tan 47° 10'
4. cot 52° 30'
5. sec 40° 30'

Find by the interpolation method the value of $\angle A$ to the nearest minute in problems 6 to 10.

6. sin $A = .0901$
7. tan $A = .3411$
8. cos $A = .4173$
9. cot $A = .8491$
10. csc $A = 1.4804$

## 14.5 Oblique Triangles

We can use right triangle methods to solve most oblique triangles by introducing perpendiculars and resolving the oblique triangle into two right triangles.

For example:

1. Triangle $ABC$ can be resolved into right triangles $ADC$ and $BDC$ by introducing the perpendicular $CD$.

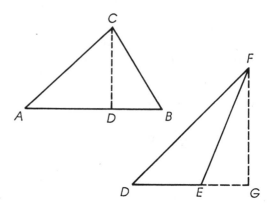

**2.** Triangle *DEF* can be resolved into right triangles *DGF* and *EGF* by extending *DE* and dropping the perpendicular *FG*.

**3.** Triangle *HJK* can be resolved into right triangles *HLJ* and *KLJ* by introducing the perpendicular *JL*.

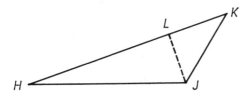

In practical problems, however, it is often impossible or too cumbersome to use a right triangle, and in such cases formulas for oblique angles are needed.

There are three important formulas that may be used in the solution of triangles of any shape. They are known as the law of sines, the law of cosines, and the law of tangents.

For our purposes it will be sufficient to state the law, give the corresponding formulas, and show the application of the law to the solution of problems involving oblique triangles.

### The Law of Sines

The sides of a triangle are proportional to the sines of their opposite angles.

$$\frac{a}{\sin A} = \frac{b}{\sin B} = \frac{c}{\sin C} \quad \text{or}$$

$$\frac{a}{b} = \frac{\sin A}{\sin B}, \frac{b}{c} = \frac{\sin B}{\sin C}, \frac{a}{c} = \frac{\sin A}{\sin C}$$

### The Law of Cosines

The square of any side of a triangle is equal to the sum of the squares of the other two sides minus twice their product times the cosine of the included angle.

$$a^2 = b^2 + c^2 - 2bc \cos A$$

$$b^2 = a^2 + c^2 - 2ac \cos B$$

$$c^2 = a^2 + b^2 - 2ab \cos C \quad \text{or}$$

$$a = \sqrt{b^2 + c^2 - 2bc \cos A}$$

$$b = \sqrt{a^2 + c^2 - 2ac \cos B}$$

$$c = \sqrt{a^2 + b^2 - 2ab \cos C}$$

### The Law of Tangents

The difference between any two sides of a triangle is to their sum as the tangent of half the difference between their opposite angles is to the tangent of half their sum.

$$\frac{a - b}{a + b} = \frac{\tan \frac{1}{2}(A - B)}{\tan \frac{1}{2}(A + B)}$$

$$\frac{a - c}{a + c} = \frac{\tan \frac{1}{2}(A - C)}{\tan \frac{1}{2}(A + C)}$$

$$\frac{b - c}{b + c} = \frac{\tan \frac{1}{2}(B - C)}{\tan \frac{1}{2}(B + C)}$$

or if $b > a$, then

$$\frac{b - a}{b + a} = \frac{\tan \frac{1}{2}(B - A)}{\tan \frac{1}{2}(B + A)}$$

Any triangle has six parts, namely, three angles and the sides opposite the angles.

In order to solve a triangle three independent parts must be known in addition to the fact that the sum of the angles of any triangle equals 180°.

In problems involving triangles there occur the following four combinations of parts which if known will determine the size and form of the triangle.

I. One side and two angles are known.

II. Two sides and the included angle are known.

III. Three sides are known.

IV. Two sides and the angle opposite one of them are known.

### Applying the Laws of Sine, Tangent, and Cosine to Oblique Triangles

**Case I:** One side and two angles are known

EXAMPLE 1: Given $\angle A = 56°$, $\angle B = 69°$ and $a = 467$, find $b$ and $c$.

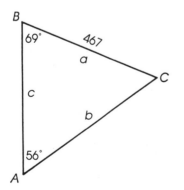

SOLUTION: We use the law of sines.

Formulas needed:

   **1.** $C = 180° - (\angle A + \angle B)$

**2.** $\dfrac{b}{a} = \dfrac{\sin B}{\sin A}$

Therefore $b = \dfrac{a \sin B}{\sin A}$

**3.** $\dfrac{c}{a} = \dfrac{\sin C}{\sin A}$

Therefore $c = \dfrac{a \sin C}{\sin A}$

Substitute

   **1.** $\angle C = 180° - (56° + 69°) = 55°$

   **2.** $b = \dfrac{467 \times .9336}{.8290} = 525.9$

   **3.** $c = \dfrac{467 \times .8192}{.8290} = 461.5$

**Case II:** Two sides and the included angle are known

EXAMPLE 2: Given $a = 17$, $b = 12$, and $\angle C = 58°$, find $\angle A$, $\angle B$, and $c$.

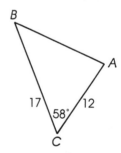

SOLUTION: We use the law of tangents to obtain $\angle A$ and $\angle B$ and the law of sines to obtain $c$.

Formulas needed:

   **1.** $A + B = 180° - C$ and $\frac{1}{2}(A + B) = \frac{1}{2}(180° - C)$

When $\frac{1}{2}(A + B)$ has been determined $\frac{1}{2}(A - B)$ is found by the following:

**2.** $\dfrac{a - b}{a + b} = \dfrac{\tan \frac{1}{2}(A - B)}{\tan \frac{1}{2}(A + B)}$

Therefore $\tan \frac{1}{2}(A - B) =$

$\dfrac{a - b}{a + b} \times \tan \frac{1}{2}(A + B)$

**3.** $\angle A = \frac{1}{2}(A + B) + \frac{1}{2}(A - B)$
in which the $B$'s cancel out

**4.** $\angle B = \frac{1}{2}(A + B) - \frac{1}{2}(A - B)$
in which the $A$'s cancel out

**5.** $\dfrac{c}{a} = \dfrac{\sin C}{\sin A}$

Therefore $c = \dfrac{a \sin C}{\sin A}$

Substitute

**1.** $\frac{1}{2}(A + B) = \frac{1}{2}(180° - 58°)$
$= 61°$

**2.** $\tan \frac{1}{2}(A - B) = \dfrac{17 - 12}{17 + 12} \times$
$\tan 61° = .3110$

which is the tan of 17° 16′ and equal to
$\frac{1}{2}(A - B)$

**3.** $\angle A = 61° + 17° \, 16′ = 78° \, 16′$

**4.** $\angle B = 61° - 17° \, 16′ = 43° \, 44′$

**5.** $c = \dfrac{17 \times \sin 58°}{\sin 78° \, 16′} = 14.7$

This example could also be solved by the use of the law of cosines by first finding $c$ ($c = \sqrt{a^2 + b^2 - 2ab \cos C}$). When the three sides and $\angle C$ are known the law of sines can be employed to find $\angle A$ and $\angle B$. For purposes of a check, do this example by the second method.

**Case III:** Three sides are known

EXAMPLE 3: Given $a = 5$, $b = 6$, and $c = 7$, find $\angle A$, $\angle B$, and $\angle C$.

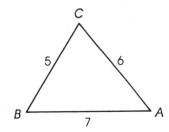

SOLUTION: We use the law of cosines and the law of sines.

Formulas needed:

**1.** $a^2 = b^2 + c^2 - 2bc \cos A$

Therefore $\cos A = \dfrac{b^2 + c^2 - a^2}{2bc}$

**2.** $\dfrac{a}{b} = \dfrac{\sin A}{\sin B}$

Therefore $\sin B = \dfrac{b \sin A}{a}$

**3.** $\angle C = 180° - (A + B)$

Substitute

$\cos A = \dfrac{36 + 49 - 25}{2(6 \times 7)} = .7143$

which is the cos of 44° 25′

$\sin B = \dfrac{6 \times .69995}{5} = .8399$

which is the sin of 57° 7′

$\angle C = 180° - (44° \, 25′ + 57° \, 7′)$
$= 78° \, 28′$

$\angle A = 44° \, 25′$

$\angle B = 57° \, 7′$

$\angle C = 78° \, 28′$

**Case IV** (*the ambiguous case*): Two sides and the angle opposite one of them are known

When given two sides of a triangle and the angle opposite one of them, there is often a possibility of two solutions unless one of the solutions is excluded by the statement of the problem.

This fact may be clarified by the next figure. It will be seen in the triangle *ABC*

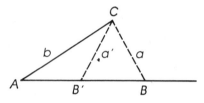

that if $\angle A$ and sides $a$ and $b$ are given, either of the triangles *ABC* or *AB'C* meet the given conditions.

By varying the relative lengths of $a$ and $b$ and the magnitude of $\angle A$, the following possibilities can be recognized.

If $a > b$, $\angle A > \angle B$, which makes $\angle B$ less than 90°, and allows for only one solution.

If $a = b$, $\angle A = \angle B$; both angles are less than 90° and only an isosceles triangle can be formed.

If $a < b$ and $\angle A$ is acute, two triangles are possible.

If $a = b \sin A$, the figure is a right triangle and only one solution is possible.

If $a < b \sin A$, no triangle is possible.

Before doing a problem of this type we can generally determine the number of possible solutions by making an approximate small-scale drawing of the given parts.

In the cases where there are two possible solutions and the unknown parts are $\angle B$, $\angle C$, and side $c$, the second set of unknown

parts should be designated as $\angle B'$, $\angle C'$, and side $c'$. They will then be found as follows:

$B' = 180° - B$, because when an angle is determined by its sine, it has two possible values that are supplementary to each other.

$$C' = 180° - (A + B')$$

$$c' = \frac{a \sin C'}{\sin A}$$

EXAMPLE 4: Given $a = 5$, $b = 8$, and $\angle A = 30°$, find $\angle B$, $\angle C$, and side $c$.

SOLUTION: Here $a < b$ and $\angle A$ is acute. Therefore two triangles are possible.

Formulas needed for $\triangle ABC$:

1. $\dfrac{b}{a} = \dfrac{\sin B}{\sin A}$

   Therefore $\sin B = \dfrac{b \sin A}{a}$

2. $\angle C = 180° - (A + B)$

3. $\dfrac{c}{a} = \dfrac{\sin C}{\sin A}$

   Therefore $c = \dfrac{a \sin C}{\sin A}$

Substitute

1. $\sin B = \dfrac{8 \times .5000}{5} = .8000$

which is the sin of 53° 8'

2. $\angle C = 180° - (30° + 53° 8') = 96° 52'$

**3.** $c = \dfrac{5 \times .9928}{.5000} = 9.928$

$$\angle B = 53° 8'$$

$$\angle C = 96° 52'$$

$$c = 9.928$$

To find $\angle B'$, $\angle C'$, and $c'$:

$$\angle B' = 180° - B = 126° 52'$$

$$\angle C' = 180° - (A + B') = 23° 8'$$

## Exercise Set 14.5

In working out the problems in this exercise apply the principles for solving oblique triangles.

1. Given $\angle A = 45°$, $\angle B = 60°$ and $c = 9.562$, find $a$ and $b$.
2. Given $a = 43$, $\angle A = 43°$, and $\angle B = 68°$, find $\angle C$, $b$, and $c$.
3. Given $a = 22$, $b = 13$, and $\angle C = 68°$, find $\angle A$, $\angle B$, and $c$.
4. Given $a = 27$, $b = 26$, $c = 34$, find $\angle A$, $\angle B$, and $\angle C$.
5. Given $a = 8$, $b = 5$, and $\angle A = 21$, find $c$, $\angle B$, and $\angle C$.

## Chapter 14 Glossary

**Cosecant**   The constant ratio $c/a$, of $\angle A$.
**Cosine**   The constant ratio $b/c$, of $\angle A$.
**Cotangent**   The constant ratio $b/a$, of $\angle A$.
**Function**   A variable quantity that depends upon another quantity for its value is called a function of the latter value.

**Law of Cosines**   The square of any side of a triangle is equal to the sum of the squares of the other two sides minus twice their product times the cosine of the included angle.
**Law of Sines**   The sides of a triangle are proportional to the sines of their opposite angles.
**Law of Tangents**   The difference between any two sides of a triangle is to their sum as the tangent of half the difference between their opposite angles is to the tangent of half their sum.
**Secant**   The constant ratio $c/b$, of $\angle A$.
**Sine**   The constant ratio $a/c$, of $\angle A$.
**Tangent**   The constant ratio $a/b$, of $\angle A$.
**Trigonometry**   The branch of mathematics that deals with the measurement of triangles.

## Chapter 14 Test

For each problem, five answers are given. Only one answer is correct. After you solve each problem, check the answer that agrees with your solution.

1. An airplane is 405 feet above a landing field when the pilot cuts out his motor. He glides to a landing at an angle of 13° with the field. How far will he glide in reaching the field?

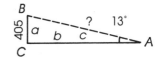

   A) 300 feet
   B) 1,248 feet
   C) 1,800 feet
   D) 1,641 feet
   E) 700 feet

**2.** An ascension balloon is moored by a rope 150 feet long. A wind blowing in an easterly direction keeps the rope taut and causes it to make an angle of 50° with the ground. What is the vertical height of the balloon from the ground?

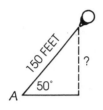

**A)** 180 feet
**B)** 114.9 feet
**C)** 177.5 feet
**D)** 189.4 feet
**E)** 126.4 feet

**3.** A carpenter has to build a ramp to be used as a loading platform for a carrier airplane. The height of the loading door is 12 feet, and the required slope or gradient of the ramp is to be 18°. How long must the ramp be?

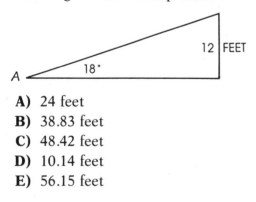

**A)** 24 feet
**B)** 38.83 feet
**C)** 48.42 feet
**D)** 10.14 feet
**E)** 56.15 feet

**4.** The fire department has a new 200-foot ladder. The greatest angle at which it can be placed against a building with safety is at 71° with the ground. What is the maximum vertical height that the ladder can reach?

**A)** 189.1 feet
**B)** 209.4 feet
**C)** 300 feet
**D)** 162.3 feet
**E)** 275 feet

**5.** A road running from the bottom of a hill to the top is 625 feet long. If the hill is 54½ feet high, what is the angle of elevation of the road?

**A)** 25°
**B)** 15°
**C)** 5°
**D)** 2°
**E)** 45°

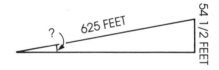

**6.** A carpenter has to build a triangular roof to a house. The roof is to be 30 feet wide. If the rafters are 17 feet long, at what angle will the rafters be laid at the eaves?

**A)** 34°
**B)** 19° 30′
**C)** 28° 05′
**D)** 42° 10′
**E)** 25°

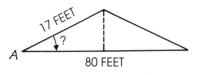

**7.** Desiring to measure distance across a pond, a surveyor standing at point *A* sighted on a point *B* across the pond. From *A* he ran a line *AC*, making an angle of 27° with *AB*. From *B* he ran a line perpendicular to *AC*. He measured the line *AC* to be 681 feet. What is the distance across the pond from *A* to *B?*

**A)** 100 feet
**B)** 764.3 feet
**C)** 681 feet
**D)** 862.8 feet
**E)** 275 feet

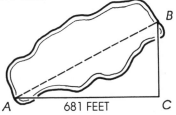

**8.** A scout on a hill 125 feet above a lake sights a boat on the water at an angle of depression of 10° as shown. What is the exact distance from the scout to the boat?

**A)** 240.5 feet
**B)** 720 feet
**C)** 468.4 feet
**D)** 1020 feet
**E)** 520 feet

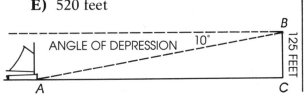

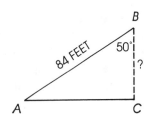

**9.** A mountain climber stretches a cord from the rocky ledge of a sheer cliff to a point on a horizontal plane, making an angle of 50° with the ledge. The cord is 84 feet long. What is the vertical height of the rocky ledge from its base?

**A)** 45 feet
**B)** 82 feet
**C)** 76.8 feet
**D)** 54 feet
**E)** 66.2 feet

**10.** A 100-foot ladder is placed against the side of a house with the foot of the ladder 16½ feet away from the building. What angle does the ladder make with the ground?

**A)** 65°
**B)** 25° 40′
**C)** 80° 30′
**D)** 72° 20′
**E)** 45°

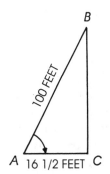

**11.** An engineer desires to learn the height of a cone-shaped hill. He measures its diameter to be 280 feet. From a point on the circumference of the base he determines that the angle of elevation is 43°. What is the altitude?

A) 130.55 feet
B) 260 feet
C) 125.45 feet
D) 560 feet
E) 290 feet

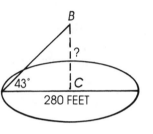

**12.** From a lookout tower 240 feet high an enemy tank division is sighted at an angle of depression which is measured to be 10°. How far is the enemy away from the lookout tower if they are both on the same level?

A) 1,361.11 feet
B) 642.25 feet
C) 866 feet
D) 2,434.16 feet
E) 987.14 feet

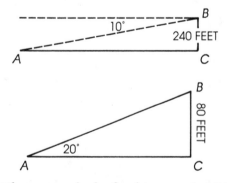

**13.** The upper deck of a ship stands 30 feet above the level of its dock. A runway to the deck is to be built having an angle of inclination of 20°. How far from the boat should it start?

A) 60 feet
B) 76.25 feet
C) 82.42 feet
D) 42.30 feet
E) 45.7 feet

**14.** From a boathouse 100 feet above the level of a lake two rowing crews were sighted racing in the direction of the boathouse. The boats were directly in a line with each other. The leading boat was sighted at an angle of depression equal to 15°, and the other at 14°. How far apart were the boats?

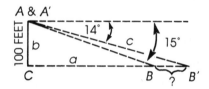

A) 373.21 feet
B) 27.87 feet
C) 64.14 feet
D) 401.08 feet
E) 174.12 feet

**15.** A clock on the tower of a building is observed from two points which are on the same level and in the same straight line with the foot of the tower. At the nearer point the angle of elevation to the clock is 60°, and at the farther point it is 30°. If the two points are 300 feet apart, what is the height of the clock?

A) 130.8 feet
B) 400 feet
C) 259.8 feet
D) 360.4 feet
E) 250 feet

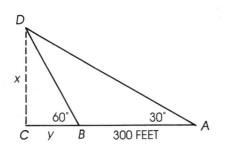

**16.** Two airplane spotters, Mary and Jake, are 1.83 miles apart on the same level of ground. Jake is due north of Mary. At the same instant they both spot an airplane to the north, which makes an angle of elevation of 67° 31′ for Mary and 82° 16′ for Jake. What is the altitude of the airplane from the ground?

A) 2.5 miles
B) 6.6 miles
C) 4 miles
D) 3.2 miles
E) 5.2 miles

**17.** An observer on a boat anchored offshore sights on two points, *A* and *B*, on the shore. He determines the distance from himself to point *A* to be 985 feet, and the distance between *A* and *B* as 1,460 feet. The angle to the observer subtended by the points on shore is 64° 20′. How far is it from the observer to point *B*?

A) 1,585.6 feet
B) 1,242.6 feet
C) 1,760 feet
D) 927.7 feet
E) 1,340 feet

**18.** An observer at a fire tower spots a fire in a forest area extending across a stretch of land from the lake to the ridge. The distance from the tower to the lake is 5 miles, and to the ridge, 5½ miles. The angle subtended by the stretch of land to the tower is 50°. What is the distance across which the fire extends?

A) 6 miles
B) 4.46 miles
C) 3.42 miles
D) 8.5 miles
E) 5.7 miles

**19.** Two Boy Scouts start from camp at the same time and branch out at an angle of 33° to each other. If one Scout travels at the rate of 1 mile per hour while the other travels at the rate of 3 miles per hour, how far apart will they be at the end of 2 hours? (Hint: Solve by cosine law.)

A) 3 miles
B) 5.42 miles
C) 4.46 miles
D) 8.56 miles
E) 7.4 miles

**20.** A cannon is placed in position at point *A* to fire upon an enemy fort located on a mountain. The airline distance from the gun to the fort has been determined as 5 miles. The distance on a horizontal plane from the gun to a point *C* at the base of the mountain is 3½ miles. From this point at the base to the fort itself the distance is 1.8 miles. What is the angle of depression from the fort to the cannon?

A) 38° 59′
B) 27° 21′
C) 22° 16′
D) 13° 40′
E) 41°

# Appendices

## A: Final Test

For each problem, five answers are given. Only one answer is correct. After you solve each problem, check the answer that agrees with your solution.

1. A brick wall falls over, leaving twenty-four inches standing. If $\frac{5}{8}$ of the wall has fallen, how many inches of brick are needed to rebuild the wall?

   A) 15 inches     D) 40 inches
   B) 30 inches     E) 27 inches
   C) 36 inches

2. Two trucks together weigh 1,800 pounds. The lighter truck weighs $\frac{1}{2}$ as much as the heavier truck. What is the weight of the heavier truck?

   A) 1,000 pounds     D) 1,600 pounds
   B) 1,200 pounds     E) 800 pounds
   C) 1,400 pounds

3. 12% of Harry's money equals 18% of Gerry's money. The poorer of the two has $300. How much does the richer of the two have?

   A) $350     D) $450
   B) $374     E) $500
   C) $400

4. How much would you have to lend for $1\frac{1}{2}$ years at 4% in order to get $424 back?

   A) $328.56     D) $390
   B) $407.04     E) $378.33
   C) $400

**5.** Marty's Shoe Shop wishes to make a 20% profit on shoes. At what price must he buy them in order to sell them at $4.50 a pair?

A) $2.50      D) $3.75
B) $4.25      E) $2.75
C) $3.50

**6.** A supply pipe with a capacity of 8 gallons per minute can fill a reservoir in 18 hours. What capacity pipe would be needed to fill the reservoir in 10 hours?

A) $4\frac{4}{9}$      D) 24
B) $14\frac{2}{5}$      E) $17\frac{1}{3}$
C) $22\frac{1}{2}$

**7.** The formula for the length of the sides of a right triangle is $c^2 = a^2 + b^2$. If $c = 15$, and $a = 12$, what does $b$ equal?

A) 9      D) 12
B) 10      E) 17
C) 11

**8.** What will be the diameter of a wheel whose area is 264 square inches?

A) 15.25 inches
B) 18.33 inches
C) 21.66 inches
D) 24.50 inches
E) 13.56 inches

**9.** The radius of a circular room having a tile floor is 21 feet. Sherry wishes to use the tiling in a rectangular room with the same area that is 14 feet wide. What will the length of the rectangular room be?

A) 21 feet      D) 120 feet
B) 63 feet      E) 75 feet
C) 99 feet

**10.** Alan has six times as many nickels as he has dimes. The value of his total money is $4.80. How many nickels does he have?

A) 60      D) 80
B) 68      E) 55
C) 72

**11.** What is the cube root of 262,144?

A) 512      D) 32
B) 64      E) 28
C) 256

**12.** What is the capacity of a can 4 inches in diameter and 6 inches high?

A) $301\frac{5}{7}$ cubic inches
B) 150.11 cubic inches
C) 75.36 cubic inches
D) 13.5 cubic inches
E) 98 cubic inches

**13.** Which of the following is one of the factors of $15x^2 - 21xy - 18y^2$?

A) $5x - 3y$      D) $3x - 6y$
B) $5x + 6y$      E) $6x + 3y$
C) $3x - 3y$

**14.** What is the sum of the angles of a hexagon?

A) 360°      D) 180°
B) 720°      E) 840°
C) 540°

# B: Table of Square Roots

| NUMBER $n$ | SQUARE $n^2$ | SQUARE ROOT $\sqrt{n}$ | NUMBER $n$ | SQUARE $n^2$ | SQUARE ROOT $\sqrt{n}$ |
|---|---|---|---|---|---|
| 1 | 1 | 1.000 | 51 | 2601 | 7.141 |
| 2 | 4 | 1.414 | 52 | 2704 | 7.211 |
| 3 | 9 | 1.732 | 53 | 2809 | 7.280 |
| 4 | 16 | 2.000 | 54 | 2916 | 7.348 |
| 5 | 25 | 2.236 | 55 | 3025 | 7.416 |
| 6 | 36 | 2.449 | 56 | 3136 | 7.483 |
| 7 | 49 | 2.645 | 57 | 3249 | 7.549 |
| 8 | 64 | 2.828 | 58 | 3364 | 7.615 |
| 9 | 81 | 3.000 | 59 | 3481 | 7.681 |
| 10 | 100 | 3.162 | 60 | 3600 | 7.746 |
| 11 | 121 | 3.316 | 61 | 3721 | 7.810 |
| 12 | 144 | 3.464 | 62 | 3844 | 7.874 |
| 13 | 169 | 3.605 | 63 | 3969 | 7.937 |
| 14 | 196 | 3.741 | 64 | 4096 | 8.000 |
| 15 | 225 | 3.873 | 65 | 4225 | 8.062 |
| 16 | 256 | 4.000 | 66 | 4356 | 8.124 |
| 17 | 289 | 4.123 | 67 | 4489 | 8.185 |
| 18 | 324 | 4.242 | 68 | 4624 | 8.246 |
| 19 | 361 | 4.358 | 69 | 4761 | 8.306 |
| 20 | 400 | 4.472 | 70 | 4900 | 8.366 |
| 21 | 441 | 4.582 | 71 | 5041 | 8.426 |
| 22 | 484 | 4.690 | 72 | 5184 | 8.485 |
| 23 | 529 | 4.795 | 73 | 5329 | 8.544 |
| 24 | 576 | 4.899 | 74 | 5476 | 8.602 |
| 25 | 625 | 5.000 | 75 | 5625 | 8.660 |
| 26 | 676 | 5.099 | 76 | 5776 | 8.717 |
| 27 | 729 | 5.196 | 77 | 5929 | 8.775 |
| 28 | 784 | 5.291 | 78 | 6084 | 8.831 |
| 29 | 841 | 5.385 | 79 | 6241 | 8.888 |
| 30 | 900 | 5.477 | 80 | 6400 | 8.944 |
| 31 | 961 | 5.567 | 81 | 6561 | 9.000 |
| 32 | 1024 | 5.656 | 82 | 6724 | 9.055 |
| 33 | 1089 | 5.744 | 83 | 6889 | 9.110 |
| 34 | 1156 | 5.831 | 84 | 7056 | 9.165 |
| 35 | 1225 | 5.916 | 85 | 7225 | 9.219 |
| 36 | 1296 | 6.000 | 86 | 7396 | 9.273 |
| 37 | 1369 | 6.082 | 87 | 7569 | 9.327 |
| 38 | 1444 | 6.164 | 88 | 7744 | 9.380 |
| 39 | 1521 | 6.245 | 89 | 7921 | 9.434 |

| NUMBER n | SQUARE n² | SQUARE ROOT √n | | NUMBER n | SQUARE n² | SQUARE ROOT √n |
|---|---|---|---|---|---|---|
| 40 | 1600 | 6.324 | | 90 | 8100 | 9.486 |
| 41 | 1681 | 6.403 | | 91 | 8281 | 9.539 |
| 42 | 1764 | 6.480 | | 92 | 8464 | 9.591 |
| 43 | 1849 | 6.557 | | 93 | 8649 | 9.643 |
| 44 | 1936 | 6.633 | | 94 | 8836 | 9.695 |
| 45 | 2025 | 6.708 | | 95 | 9025 | 9.746 |
| 46 | 2116 | 6.782 | | 96 | 9216 | 9.798 |
| 47 | 2209 | 6.855 | | 97 | 9409 | 9.848 |
| 48 | 2304 | 6.928 | | 98 | 9604 | 9.899 |
| 49 | 2401 | 7.000 | | 99 | 9801 | 9.949 |
| 50 | 2500 | 7.071 | | 100 | 10000 | 10.000 |

## C: Other Multiplication Tables

| NUMBER | ×13 | ×14 | ×15 | ×16 | ×19 | ×21 | ×24 | ×25 |
|---|---|---|---|---|---|---|---|---|
| 1 | 13 | 14 | 15 | 16 | 19 | 21 | 24 | 25 |
| 2 | 26 | 28 | 30 | 32 | 38 | 42 | 48 | 50 |
| 3 | 39 | 42 | 45 | 48 | 57 | 63 | 72 | 75 |
| 4 | 52 | 56 | 60 | 64 | 76 | 84 | 96 | 100 |
| 5 | 65 | 70 | 75 | 80 | 95 | 105 | 120 | 125 |
| 6 | 78 | 84 | 90 | 96 | 114 | 126 | 144 | 150 |
| 7 | 91 | 98 | 105 | 112 | 133 | 147 | 168 | 175 |
| 8 | 104 | 112 | 120 | 128 | 152 | 168 | 192 | 200 |
| 9 | 117 | 126 | 135 | 144 | 171 | 189 | 216 | 225 |
| 10 | 130 | 140 | 150 | 160 | 190 | 210 | 240 | 250 |
| 11 | 143 | 154 | 165 | 176 | 209 | 231 | 264 | 275 |
| 12 | 156 | 168 | 180 | 192 | 228 | 252 | 288 | 300 |
| 13 | 169 | 182 | 195 | 208 | 247 | 273 | 312 | 325 |
| 14 | 182 | 196 | 210 | 224 | 266 | 294 | 336 | 350 |
| 15 | 195 | 210 | 225 | 240 | 285 | 315 | 360 | 375 |
| 16 | 208 | 224 | 240 | 256 | 304 | 336 | 384 | 400 |
| 17 | 221 | 238 | 255 | 272 | 323 | 357 | 408 | 425 |
| 18 | 234 | 252 | 270 | 288 | 342 | 378 | 432 | 450 |
| 19 | 247 | 266 | 285 | 304 | 361 | 399 | 456 | 475 |
| 20 | 260 | 280 | 300 | 320 | 380 | 420 | 480 | 500 |
| 21 | 273 | 294 | 315 | 336 | 399 | 441 | 504 | 525 |
| 22 | 286 | 308 | 330 | 352 | 418 | 462 | 528 | 550 |
| 23 | 299 | 322 | 345 | 368 | 437 | 483 | 552 | 575 |
| 24 | 312 | 336 | 360 | 384 | 456 | 504 | 576 | 600 |
| 25 | 325 | 350 | 375 | 400 | 475 | 525 | 600 | 625 |

# D: Tables of Measures

## a) Length, or Linear Measure

### U.S. SYSTEM

12 inches (in. or ″) = 1 foot (ft. or ′)
3 feet or 36 inches = 1 yard (yd.)
1,760 yards = 1 mile
5,280 feet = 1 mile

### METRIC SYSTEM

| UNIT | | METERS |
|---|---|---|
| 1 millimeter (mm) = | | 0.001 |
| 10 millimeters = 1 centimeter (cm) | = | 0.01 |
| 10 centimeters = 1 decimeter (dm) | = | 0.1 |
| 10 decimeters = 1 meter (m) | = | 1. |
| 10 meters = 1 dekameter (dam) | = | 10. |
| 10 dekameters = 1 hectometer (hm) | = | 100. |
| 10 hectometers = 1 kilometer (km) | = | 1,000. |

## b) Area, or Square Measure

### U.S. SYSTEM

144 square inches = 1 square foot (sq. ft.)
9 square feet = 1 square yard (sq. yd.)
4,840 square yards = 1 acre (A.)
640 acres = 1 square mile

### METRIC SYSTEM

100 square millimeters (sq mm) = 1 square centimeter (sq cm)
100 square centimeters = 1 square decimeter (sq dm)
100 square decimeters = 1 square meter (sq m)
100 square meters = 1 square dekameter (sq dm) or are (a)
100 square ares = 1 square hectometer (sq hm) or hectare (h)
100 square hectares = 1 square kilometer (sq km)

## c) Volume, or Cubic Measure

### U.S. SYSTEM

1,728 cubic inches (cu. in.) = 1 cubic foot (cu. ft.)
27 cubic feet = 1 cubic yard (cu. yd.)

### LIQUID MEASURE

4 gills (gi.) = 1 pint (pt.)
2 pints = 1 quart (qt.)
4 quarts = 1 gallon (gal.)

## METRIC SYSTEM

1,000 cubic millimeters ($mm^3$) = 1 cubic centimeter ($cm^3$)
1,000 cubic centimeters      = 1 cubic decimeter ($dm^3$)
1,000 cubic decimeters      = 1 cubic meter ($m^3$)

## LIQUID MEASURE

10 milliliters (ml)   = 1 centiliter (cl)
10 centiliters (cl)   = 1 deciliter (dl)
10 deciliters      = 1 liter (l) = 1 cubic meter
10 cubic liters    = 1 dekaliter (dal)
10 dekaliters     = 1 hectoliter (hl)

## d) Weight

Four scales of weight are used in the U.S.

a. **Troy**—for weighing gold, silver, and other precious metals.

b. **Apothecaries'**—used by druggists for weighing chemicals.

c. **Avoirdupois**—used for all general purposes.

d. **Metric**—used in scientific work.

## AVOIRDUPOIS WEIGHT

16 drams (dr.) = 1 ounce (oz.)
16 ounces     = 1 pound (lb.)
7,000 grains (gr.) = 1 pound
100 pounds     = 1 hundredweight (cwt.)
2,000 pounds   = 1 ton or short ton
112 pounds     = 1 cwt. old measure
2,240 pounds   = 1 long ton

## TROY WEIGHT

24 grains (gr.)     = 1 pennyweight (pwt.)
20 pennyweights = 1 ounce (oz.)
12 ounces       = 1 pound (lb.)
5,760 grains      = 1 pound
3.2 grains       = 1 carat (kt.)

The carat, as defined in the table, is used to weigh diamonds. The same term is used to indicate the purity of gold. In this case, a carat means a twenty-fourth part. Thus, 14 Kt. gold means that 14 parts ($^{14}\!/_{24}$) are pure gold and that 10 parts ($^{10}\!/_{24}$) are of other metals.

## APOTHECARIES' WEIGHT

20 grains (gr.) = 1 scruple
3 scruples    = 1 dram
8 drams      = 1 ounce
12 ounces     = 1 pound
5,760 grains    = 1 pound

## METRIC WEIGHT

10 milligrams (mg) = 1 centigram (cg)
10 centigrams     = 1 decigram (dg)
10 decigrams      = 1 gram (g)
10 grams         = 1 dekagram (Dg)
10 dekagrams      = 1 hectogram (hg)
10 hectograms     = 1 kilogram (kg)

## e) Time

60 seconds (sec. or ") = 1 minute (min. or ')
60 minutes         = 1 hour (hr.)
24 hours           = 1 day (da.)
7 days             = 1 week (wk.)
365 days           = 1 common year
366 days           = 1 leap year
12 calendar months = 1 year
10 years           = 1 decade
100 years          = 1 century (C.)

# E: Table of Trigonometric Functions

| DEGREES | SIN | COS | TAN | COT | SEC | CSC | |
|---|---|---|---|---|---|---|---|
| 0° 00′ | .0000 | 1.0000 | .0000 | — | 1.000 | — | 90° 00′ |
| 10 | 029 | 000 | 029 | 343.8 | 000 | 343.8 | 50 |
| 20 | 058 | 000 | 058 | 171.9 | 000 | 171.9 | 40 |
| 30 | .0087 | 1.0000 | .0087 | 114.6 | 1.000 | 114.6 | 30 |
| 40 | 116 | 9999 | 116 | 85.94 | 000 | 85.95 | 20 |
| 50 | 145 | 999 | 145 | 68.75 | 000 | 68.76 | 10 |
| 1° 00′ | .0175 | .9998 | .0175 | 57.29 | 1.000 | 57.30 | 89° 00′ |
| 10 | 204 | 998 | 204 | 49.10 | 000 | 49.11 | 50 |
| 20 | 233 | 997 | 233 | 42.96 | 000 | 42.98 | 40 |
| 30 | .0262 | .9997 | .0262 | 38.19 | 1.000 | 38.20 | 30 |
| 40 | 291 | 996 | 291 | 34.37 | 000 | 34.38 | 20 |
| 50 | 320 | 995 | 320 | 31.24 | 001 | 31.26 | 10 |
| 2° 00′ | .0349 | .9994 | .0349 | 28.64 | 1.001 | 28.65 | 88° 00′ |
| 10 | 378 | 993 | 378 | 26.43 | 001 | 26.45 | 50 |
| 20 | 407 | 992 | 407 | 24.54 | 001 | 24.56 | 40 |
| 30 | .0436 | .9990 | .0437 | 22.90 | 1.001 | 22.93 | 30 |
| 40 | 465 | 989 | 466 | 21.47 | 001 | 21.49 | 20 |
| 50 | 494 | 988 | 495 | 20.21 | 001 | 20.23 | 10 |
| 3° 00′ | .0523 | .9986 | .0524 | 19.08 | 1.001 | 19.11 | 87° 00′ |
| 10 | 552 | 985 | 553 | 18.07 | 002 | 18.10 | 50 |
| 20 | 581 | 983 | 582 | 17.17 | 002 | 17.20 | 40 |
| 30 | .0610 | .9981 | .0612 | 16.35 | 1.002 | 16.38 | 30 |
| 40 | 640 | 980 | 641 | 15.60 | 002 | 15.64 | 20 |
| 50 | 669 | 978 | 670 | 14.92 | 002 | 14.96 | 10 |
| 4° 00′ | .0698 | .9976 | .0699 | 14.30 | 1.002 | 14.34 | 86° 00′ |
| 10 | 727 | 974 | 729 | 13.73 | 003 | 13.76 | 50 |
| 20 | 756 | 971 | 758 | 13.20 | 003 | 13.23 | 40 |
| 30 | .0785 | .9969 | .0787 | 12.71 | 1.003 | 12.75 | 30 |
| 40 | 814 | 967 | 816 | 12.25 | 003 | 12.29 | 20 |
| 50 | 843 | 964 | 846 | 11.83 | 004 | 11.87 | 10 |
| 5° 00′ | .0872 | .9962 | .0875 | 11.43 | 1.004 | 11.47 | 85° 00′ |
| 10 | 901 | 959 | 904 | 11.06 | 004 | 11.10 | 50 |
| 20 | 929 | 957 | 934 | 10.71 | 004 | 10.76 | 40 |
| 30 | .0958 | .9954 | .0963 | 10.39 | 1.005 | 10.43 | 30 |
| 40 | 987 | 951 | 992 | 10.08 | 005 | 10.13 | 20 |
| 50 | .1016 | 948 | .1022 | 9.788 | 005 | 9.839 | 10 |
| | Cos | Sin | Cot | Tan | Csc | Sec | Degrees |

| DEGREES | SIN | COS | TAN | COT | SEC | CSC | |
|---|---|---|---|---|---|---|---|
| 6° 00′ | .1045 | .9945 | .1051 | 9.514 | 1.006 | 9.567 | 84° 00′ |
| 10 | 074 | 942 | 080 | 9.255 | 006 | 9.309 | 50 |
| 20 | 103 | 939 | 110 | 9.010 | 006 | 9.065 | 40 |
| 30 | .1132 | .9936 | .1139 | 8.777 | 1.006 | 8.834 | 30 |
| 40 | 161 | 932 | 169 | 8.556 | 007 | 8.614 | 20 |
| 50 | 190 | 929 | 198 | 8.345 | 007 | 8.405 | 10 |
| 7° 00′ | .1219 | .9925 | .1228 | 8.144 | 1.008 | 8.206 | 83° 00′ |
| 10 | 248 | 922 | 257 | 7.953 | 008 | 8.016 | 50 |
| 20 | 276 | 918 | 287 | 7.770 | 008 | 7.834 | 40 |
| 30 | .1305 | .9914 | .1317 | 7.596 | 1.009 | 7.661 | 30 |
| 40 | 334 | 911 | 346 | 7.429 | 009 | 7.496 | 20 |
| 50 | 363 | 907 | 376 | 7.269 | 009 | 7.337 | 10 |
| 8° 00′ | .1392 | .9903 | .1405 | 7.115 | 1.010 | 7.185 | 82° 00′ |
| 10 | 421 | 899 | 435 | 6.968 | 010 | 7.040 | 50 |
| 20 | 449 | 894 | 465 | 6.827 | 011 | 6.900 | 40 |
| 30 | .1478 | .9890 | .1495 | 6.691 | 1.011 | 6.765 | 30 |
| 40 | 507 | 886 | 524 | 6.561 | 012 | 6.636 | 20 |
| 50 | 536 | 881 | 554 | 6.435 | 012 | 6.512 | 10 |
| 9° 00′ | .1564 | .9877 | .1584 | 6.314 | 1.012 | 6.392 | 81° 00′ |
| 10 | 593 | 872 | 614 | 197 | 013 | 277 | 50 |
| 20 | 622 | 868 | 644 | 084 | 013 | 166 | 40 |
| 30 | .1650 | .9863 | .1673 | 5.976 | 1.014 | 6.059 | 30 |
| 40 | 679 | 858 | 703 | 871 | 014 | 5.955 | 20 |
| 50 | 708 | 853 | 733 | 769 | 015 | 855 | 10 |
| 10° 00′ | .1736 | .9848 | .1763 | 5.671 | 1.015 | 5.759 | 80° 00′ |
| 10 | 765 | 843 | 793 | 576 | 016 | 665 | 50 |
| 20 | 794 | 838 | 823 | 485 | 016 | 575 | 40 |
| 30 | .1822 | .9833 | .1853 | 5.396 | 1.017 | 5.487 | 30 |
| 40 | 851 | 827 | 883 | 309 | 018 | 403 | 20 |
| 50 | 880 | 822 | 914 | 226 | 018 | 320 | 10 |
| 11° 00′ | .1908 | .9816 | .1944 | 5.145 | 1.019 | 5.241 | 79° 00′ |
| 10 | 937 | 811 | 974 | 066 | 019 | 164 | 50 |
| 20 | 965 | 805 | .2004 | 4.989 | 020 | 089 | 40 |
| 30 | .1994 | .9799 | .2035 | 4.915 | 1.020 | 5.016 | 30 |
| 40 | .2022 | 793 | 065 | 843 | 021 | 4.945 | 20 |
| 50 | 051 | 787 | 095 | 773 | 022 | 876 | 10 |
| 12° 00′ | .2079 | .9781 | .2126 | 4.705 | 1.022 | 4.810 | 78° 00′ |
| 10 | 108 | 775 | 156 | 638 | 023 | 745 | 50 |
| 20 | 136 | 769 | 186 | 574 | 024 | 682 | 40 |
| 30 | .2164 | .9763 | .2217 | 4.511 | 1.024 | 4.620 | 30 |
| 40 | 193 | 757 | 247 | 449 | 025 | 560 | 20 |
| 50 | 221 | 750 | 278 | 390 | 026 | 502 | 10 |
| | Cos | Sin | Cot | Tan | Csc | Sec | Degrees |

| DEGREES | SIN | COS | TAN | COT | SEC | CSC | |
|---|---|---|---|---|---|---|---|
| 13° 00′ | .2250 | .9744 | .2309 | 4.331 | 1.026 | 4.445 | 77° 00′ |
| 10 | 278 | 737 | 339 | 275 | 027 | 390 | 50 |
| 20 | 306 | 730 | 370 | 219 | 028 | 336 | 40 |
| 30 | .2334 | .9724 | .2401 | 4.165 | 1.028 | 4.284 | 30 |
| 40 | 363 | 717 | 432 | 113 | 029 | 232 | 20 |
| 50 | 391 | 710 | 462 | 061 | 030 | 182 | 10 |
| 14° 00′ | .2419 | .9703 | .2493 | 4.011 | 1.031 | 4.134 | 76° 00′ |
| 10 | 447 | 696 | 524 | 3.962 | 031 | 086 | 50 |
| 20 | 476 | 689 | 555 | 914 | 032 | 039 | 40 |
| 30 | .2504 | .9681 | .2586 | 3.867 | 1.033 | 3.994 | 30 |
| 40 | 532 | 674 | 617 | 821 | 034 | 950 | 20 |
| 50 | 560 | 667 | 648 | 776 | 034 | 906 | 10 |
| 15° 00′ | .2588 | .9659 | .2679 | 3.732 | 1.035 | 3.864 | 75° 00′ |
| 10 | 616 | 652 | 711 | 689 | 036 | 822 | 50 |
| 20 | 644 | 644 | 742 | 647 | 037 | 782 | 40 |
| 30 | .2672 | .9636 | .2773 | 3.606 | 1.038 | 3.742 | 30 |
| 40 | 700 | 628 | 805 | 566 | 039 | 703 | 20 |
| 50 | 728 | 621 | 836 | 526 | 039 | 665 | 10 |
| 16° 00′ | .2756 | .9613 | .2867 | 3.487 | 1.040 | 3.628 | 74° 00′ |
| 10 | 784 | 605 | 899 | 450 | 041 | 592 | 50 |
| 20 | 812 | 596 | 931 | 412 | 042 | 556 | 40 |
| 30 | .2840 | .9588 | .2962 | 3.376 | 1.043 | 3.521 | 30 |
| 40 | 868 | 580 | 994 | 340 | 044 | 487 | 20 |
| 50 | 896 | 572 | .3026 | 305 | 045 | 453 | 10 |
| 17° 00′ | .2924 | .9563 | .3057 | 3.271 | 1.046 | 3.420 | 73° 00′ |
| 10 | 952 | 555 | 089 | 237 | 047 | 388 | 50 |
| 20 | 979 | 546 | 121 | 204 | 048 | 356 | 40 |
| 30 | .3007 | .9537 | .3153 | 3.172 | 1.049 | 3.326 | 30 |
| 40 | 035 | 528 | 185 | 140 | 049 | 295 | 20 |
| 50 | 062 | 520 | 217 | 108 | 050 | 265 | 10 |
| 18° 00′ | .3090 | .9511 | .3249 | 3.078 | 1.051 | 3.236 | 72° 00′ |
| 10 | 118 | 502 | 281 | 047 | 052 | 207 | 50 |
| 20 | 145 | 492 | 314 | 018 | 053 | 179 | 40 |
| 30 | .3173 | .9483 | .3346 | 2.989 | 1.054 | 3.152 | 30 |
| 40 | 201 | 474 | 378 | 960 | 056 | 124 | 20 |
| 50 | 228 | 465 | 411 | 932 | 057 | 098 | 10 |
| 19° 00′ | .3256 | .9455 | .3443 | 2.904 | 1.058 | 3.072 | 71° 00′ |
| 10 | 283 | 446 | 476 | 877 | 059 | 046 | 50 |
| 20 | 311 | 436 | 508 | 850 | 060 | 021 | 40 |
| 30 | .3338 | .9426 | .3541 | 2.824 | 1.061 | 2.996 | 30 |
| 40 | 365 | 417 | 574 | 798 | 062 | 971 | 20 |
| 50 | 393 | 407 | 607 | 773 | 063 | 947 | 10 |
| | Cos | Sin | Cot | Tan | Csc | Sec | Degrees |

| DEGREES | SIN | COS | TAN | COT | SEC | CSC | |
|---|---|---|---|---|---|---|---|
| 20° 00' | .3420 | .9397 | .3640 | 2.747 | 1.064 | 2.924 | 70° 00' |
| 10 | 448 | 387 | 673 | 723 | 065 | 901 | 50 |
| 20 | 475 | 377 | 706 | 699 | 066 | 878 | 40 |
| 30 | .3502 | .9367 | .3739 | 2.675 | 1.068 | 2.855 | 30 |
| 40 | 529 | 356 | 772 | 651 | 069 | 833 | 20 |
| 50 | 557 | 346 | 805 | 628 | 070 | 812 | 10 |
| 21° 00' | .3584 | .9336 | .3839 | 2.605 | 1.071 | 2.790 | 69° 00' |
| 10 | 611 | 325 | 872 | 583 | 072 | 769 | 50 |
| 20 | 638 | 315 | 906 | 560 | 074 | 749 | 40 |
| 30 | .3665 | .9304 | .3939 | 2.539 | 1.075 | 2.729 | 30 |
| 40 | 692 | 293 | 973 | 517 | 076 | 709 | 20 |
| 50 | 719 | 283 | .4006 | 496 | 077 | 689 | 10 |
| 22° 00' | .3746 | .9272 | .4040 | 2.475 | 1.079 | 2.669 | 68° 00' |
| 10 | 773 | 261 | 074 | 455 | 080 | 650 | 50 |
| 20 | 800 | 250 | 108 | 434 | 081 | 632 | 40 |
| 30 | .3827 | .9239 | .4142 | 2.414 | 1.082 | 2.613 | 30 |
| 40 | 854 | 228 | 176 | 394 | 084 | 595 | 20 |
| 50 | 881 | 216 | 210 | 375 | 085 | 577 | 10 |
| 23° 00' | .3907 | .9205 | .4245 | 2.356 | 1.086 | 2.559 | 67° 00' |
| 10 | 934 | 194 | 279 | 337 | 088 | 542 | 50 |
| 20 | 961 | 182 | 314 | 318 | 089 | 525 | 40 |
| 30 | .3987 | .9171 | .4348 | 2.300 | 1.090 | 2.508 | 30 |
| 40 | .4014 | 159 | 383 | 282 | 092 | 491 | 20 |
| 50 | 041 | 147 | 417 | 264 | 093 | 475 | 10 |
| 24° 00' | .4067 | .9135 | .4452 | 2.246 | 1.095 | 2.459 | 66° 00' |
| 10 | 094 | 124 | 487 | 229 | 096 | 443 | 50 |
| 20 | 120 | 112 | 522 | 211 | 097 | 427 | 40 |
| 30 | .4147 | .9100 | .4557 | 2.194 | 1.099 | 2.411 | 30 |
| 40 | 173 | 088 | 592 | 177 | 100 | 396 | 20 |
| 50 | 200 | 075 | 628 | 161 | 102 | 381 | 10 |
| 25° 00' | .4226 | .9063 | .4663 | 2.145 | 1.103 | 2.366 | 65° 00' |
| 10 | 253 | 051 | 699 | 128 | 105 | 352 | 50 |
| 20 | 279 | 038 | 734 | 112 | 106 | 337 | 40 |
| 30 | .4305 | .9026 | .4770 | 2.097 | 1.108 | 2.323 | 30 |
| 40 | 331 | 013 | 806 | 081 | 109 | 309 | 20 |
| 50 | 358 | 001 | 841 | 066 | 111 | 295 | 10 |
| 26° 00' | .4384 | .8988 | .4877 | 2.050 | 1.113 | 2.281 | 64° 00' |
| 10 | 410 | 975 | 913 | 035 | 114 | 268 | 50 |
| 20 | 436 | 962 | 950 | 020 | 116 | 254 | 40 |
| 30 | .4462 | .8949 | .4986 | 2.006 | 1.117 | 2.241 | 30 |
| 40 | 488 | 936 | .5022 | 1.991 | 119 | 228 | 20 |
| 50 | 514 | 923 | 059 | 977 | 121 | 215 | 10 |
| | Cos | Sin | Cot | Tan | Csc | Sec | Degrees |

| DEGREES | SIN | COS | TAN | COT | SEC | CSC | |
|---|---|---|---|---|---|---|---|
| 27° 00′ | .4540 | .8910 | .5095 | 1.963 | 1.122 | 2.203 | 63° 00′ |
| 10 | 566 | 897 | 132 | 949 | 124 | 190 | 50 |
| 20 | 592 | 884 | 169 | 935 | 126 | 178 | 40 |
| 30 | .4617 | .8870 | .5206 | 1.921 | 1.127 | 2.166 | 30 |
| 40 | 643 | 857 | 243 | 907 | 129 | 154 | 20 |
| 50 | 669 | 843 | 280 | 894 | 131 | 142 | 10 |
| 28° 00′ | .4695 | .8829 | .5317 | 1.881 | 1.133 | 2.130 | 62° 00′ |
| 10 | 720 | 816 | 354 | 868 | 134 | 118 | 50 |
| 20 | 746 | 802 | 392 | 855 | 136 | 107 | 40 |
| 30 | .4772 | .8788 | .5430 | 1.842 | 1.138 | 2.096 | 30 |
| 40 | 797 | 774 | 467 | 829 | 140 | 085 | 20 |
| 50 | 823 | 760 | 505 | 816 | 142 | 074 | 10 |
| 29° 00′ | .4848 | .8746 | .5543 | 1.804 | 1.143 | 2.063 | 61° 00′ |
| 10 | 874 | 732 | 581 | 792 | 145 | 052 | 50 |
| 20 | 899 | 718 | 619 | 780 | 147 | 041 | 40 |
| 30 | .4924 | .8704 | .5658 | 1.767 | 1.149 | 2.031 | 30 |
| 40 | 950 | 689 | 696 | 756 | 151 | 020 | 20 |
| 50 | 975 | 675 | 735 | 744 | 153 | 010 | 10 |
| 30° 00′ | .5000 | .8660 | .5774 | 1.732 | 1.155 | 2.000 | 60° 00′ |
| 10 | 025 | 646 | 812 | 720 | 157 | 1.990 | 50 |
| 20 | 050 | 631 | 851 | 709 | 159 | 980 | 40 |
| 30 | .5075 | .8616 | .5890 | 1.698 | 1.161 | 1.970 | 30 |
| 40 | 100 | 601 | 930 | 686 | 163 | 961 | 20 |
| 50 | 125 | 587 | 969 | 675 | 165 | 951 | 10 |
| 31° 00′ | .5150 | .8572 | .6009 | 1.664 | 1.167 | 1.942 | 59° 00′ |
| 10 | 175 | 557 | 048 | 653 | 169 | 932 | 50 |
| 20 | 200 | 542 | 088 | 643 | 171 | 923 | 40 |
| 30 | .5225 | .8526 | .6128 | 1.632 | 1.173 | 1.914 | 30 |
| 40 | 250 | 511 | 168 | 621 | 175 | 905 | 20 |
| 50 | 275 | 496 | 208 | 611 | 177 | 896 | 10 |
| 32° 00′ | .5299 | .8480 | .6249 | 1.600 | 1.179 | 1.887 | 58° 00′ |
| 10 | 324 | 465 | 289 | 590 | 181 | 878 | 50 |
| 20 | 348 | 450 | 330 | 580 | 184 | 870 | 40 |
| 30 | .5373 | .8434 | .6371 | 1.570 | 1.186 | 1.861 | 30 |
| 40 | 398 | 418 | 412 | 560 | 188 | 853 | 20 |
| 50 | 422 | 403 | 453 | 550 | 190 | 844 | 10 |
| 33° 00′ | .5446 | .8387 | .6494 | 1.540 | 1.192 | 1.836 | 57° 00′ |
| 10 | 471 | 371 | 536 | 530 | 195 | 828 | 50 |
| 20 | 495 | 355 | 577 | 520 | 197 | 820 | 40 |
| 30 | .5519 | .8339 | .6619 | 1.511 | 1.199 | 1.812 | 30 |
| 40 | 544 | 323 | 661 | 501 | 202 | 804 | 20 |
| 50 | 568 | 307 | 703 | 1.492 | 204 | 796 | 10 |
| | Cos | Sin | Cot | Tan | Csc | Sec | Degrees |

| DEGREES | SIN | COS | TAN | COT | SEC | CSC | |
|---|---|---|---|---|---|---|---|
| 34° 00' | .5592 | .8290 | .6745 | 1.483 | 1.206 | 1.788 | 56° 00' |
| 10 | 616 | 274 | 787 | 473 | 209 | 781 | 50 |
| 20 | 640 | 258 | 830 | 464 | 211 | 773 | 40 |
| 30 | .5664 | .8241 | .6873 | 1.455 | 1.213 | 1.766 | 30 |
| 40 | 688 | 225 | 916 | 446 | 216 | 758 | 20 |
| 50 | 712 | 208 | 959 | 437 | 218 | 751 | 10 |
| 35° 00' | .5736 | .8192 | .7002 | 1.428 | 1.221 | 1.743 | 55° 00' |
| 10 | 760 | 175 | 046 | 419 | 223 | 736 | 50 |
| 20 | 783 | 158 | 089 | 411 | 226 | 729 | 40 |
| 30 | .5807 | .8141 | .7133 | 1.402 | 1.228 | 1.722 | 30 |
| 40 | 831 | 124 | 177 | 393 | 231 | 715 | 20 |
| 50 | 854 | 107 | 221 | 385 | 233 | 708 | 10 |
| 36° 00' | .5878 | .8090 | .7265 | 1.376 | 1.236 | 1.701 | 54° 00' |
| 10 | 901 | 073 | 310 | 368 | 239 | 695 | 50 |
| 20 | 925 | 056 | 355 | 360 | 241 | 688 | 40 |
| 30 | .5948 | .8039 | .7400 | 1.351 | 1.244 | 1.681 | 30 |
| 40 | 972 | 021 | 445 | 343 | 247 | 675 | 20 |
| 50 | 995 | 004 | 490 | 335 | 249 | 668 | 10 |
| 37° 00' | .6018 | .7986 | .7536 | 1.327 | 1.252 | 1.662 | 53° 00' |
| 10 | 041 | 969 | 581 | 319 | 255 | 655 | 50 |
| 20 | 065 | 951 | 627 | 311 | 258 | 649 | 40 |
| 30 | .6088 | .7934 | .7673 | 1.303 | 1.260 | 1.643 | 30 |
| 40 | 111 | 916 | 720 | 295 | 263 | 636 | 20 |
| 50 | 134 | 898 | 766 | 288 | 266 | 630 | 10 |
| 38° 00' | .6157 | .7880 | .7813 | 1.280 | 1.269 | 1.624 | 52° 00' |
| 10 | 180 | 862 | 860 | 272 | 272 | 618 | 50 |
| 20 | 202 | 844 | 907 | 265 | 275 | 612 | 40 |
| 30 | .6225 | .7826 | .7954 | 1.257 | 1.278 | 1.606 | 30 |
| 40 | 248 | 808 | .8002 | 250 | 281 | 601 | 20 |
| 50 | 271 | 790 | 050 | 242 | 284 | 595 | 10 |
| 39° 00' | .6293 | .7771 | .8098 | 1.235 | 1.287 | 1.589 | 51° 00' |
| 10 | 316 | 753 | 146 | 228 | 290 | 583 | 50 |
| 20 | 338 | 735 | 195 | 220 | 293 | 578 | 40 |
| 30 | .6361 | .7716 | .8243 | 1.213 | 1.296 | 1.572 | 30 |
| 40 | 383 | 698 | 292 | 206 | 299 | 567 | 20 |
| 50 | 406 | 679 | 342 | 199 | 302 | 561 | 10 |
| 40° 00' | .6428 | .7660 | .8391 | 1.192 | 1.305 | 1.556 | 50° 00' |
| 10 | 450 | 642 | 441 | 185 | 309 | 550 | 50 |
| 20 | 472 | 623 | 491 | 178 | 312 | 545 | 40 |
| 30 | .6494 | .7604 | .8541 | 1.171 | 1.315 | 1.540 | 30 |
| 40 | 517 | 585 | 591 | 164 | 318 | 535 | 20 |
| 50 | 539 | 566 | 642 | 157 | 322 | 529 | 10 |
| | Cos | Sin | Cot | Tan | Csc | Sec | Degrees |

| DEGREES | SIN | COS | TAN | COT | SEC | CSC | |
|---|---|---|---|---|---|---|---|
| 41° 00′ | .6561 | .7547 | .8693 | 1.150 | 1.325 | 1.524 | 49° 00′ |
| 10 | 583 | 528 | 744 | 144 | 328 | 519 | 50 |
| 20 | 604 | 509 | 796 | 137 | 332 | 514 | 40 |
| 30 | .6626 | .7490 | .8847 | 1.130 | 1.335 | 1.509 | 30 |
| 40 | 648 | 470 | 899 | 124 | 339 | 504 | 20 |
| 50 | 670 | 451 | 952 | 117 | 342 | 499 | 10 |
| 42° 00′ | .6691 | .7431 | .9004 | 1.111 | 1.346 | 1.494 | 48° 00′ |
| 10 | 713 | 412 | 057 | 104 | 349 | 490 | 50 |
| 20 | 734 | 392 | 110 | 098 | 353 | 485 | 40 |
| 30 | .6756 | .7373 | .9163 | 1.091 | 1.356 | 1.480 | 30 |
| 40 | 777 | 353 | 217 | 085 | 360 | 476 | 20 |
| 50 | 799 | 333 | 271 | 079 | 364 | 471 | 10 |
| 43° 00′ | .6820 | .7314 | .9325 | 1.072 | 1.367 | 1.466 | 47° 00′ |
| 10 | 841 | 294 | 380 | 066 | 371 | 462 | 50 |
| 20 | 862 | 274 | 435 | 060 | 375 | 457 | 40 |
| 30 | .6884 | .7254 | .9490 | 1.054 | 1.379 | 1.453 | 30 |
| 40 | 905 | 234 | 545 | 048 | 382 | 448 | 20 |
| 50 | 926 | 214 | 601 | 042 | 386 | 444 | 10 |
| 44° 00′ | .6947 | .7193 | .9657 | 1.036 | 1.390 | 1.440 | 46° 00′ |
| 10 | 967 | 173 | 713 | 030 | 394 | 435 | 50 |
| 20 | 988 | 153 | 770 | 024 | 398 | 431 | 40 |
| 30 | .7009 | .7133 | .9827 | 1.018 | 1.402 | 1.427 | 30 |
| 40 | 030 | 112 | 884 | 012 | 406 | 423 | 20 |
| 50 | 050 | 092 | 942 | 006 | 410 | 418 | 10 |
| 45° 00′ | .7071 | .7071 | 1.0000 | 1.000 | 1.414 | 1.414 | 45° 00′ |
| | Cos | Sin | Cot | Tan | Csc | Sec | Degrees |

# F: Answers to Practically Speaking Boxes

## Answers to Practically Speaking 1.4

1. $168
2. 6

## Answers to Practically Speaking 1.5

1. $280 more
2. $160 less
3. $651.13

## Answers to Practically Speaking 2.6

1. Multiply the amount of each ingredient by 3.
2. ¾ teaspoon cinnamon
3. 1 teaspoon nutmeg
4. 1½ cups honey

## Answer to Practically Speaking 3.5

1. $9.10

## Answer to Practically Speaking 4.5

1. 750 pounds

## Answer to Practically Speaking 5.2

**1.** 122° change in Fahrenheit degrees

## Answers to Practically Speaking 5.4

**1.** No. $16
**2.** No. $16.10
**3.** Answers will vary to these questions.

## Answers to Practically Speaking 6.2

**1.** $450
**2.** $810
**3.** $810

## Answers to Practically Speaking 6.5

**1.** The $150 dress will be less expensive after the discount.

## Answers to Practically Speaking 8.3

**1.** No
**2.** A typographical error is the most likely answer. Other answers are possible.

## Answers to Practically Speaking 10.6

**1.** 40 feet
**2.** 80 white stones

## Answer to Practically Speaking 11.2

**1.** 544 square feet of carpeting

## Answers to Practically Speaking 13.3

**1.** Since there are thirteen hearts in a fifty-two-card deck, when the first card is dealt, the chances of getting a heart are 13/52; with the second, if no heart is dealt, the chances are 13/51. As eight cards have been dealt, with no hearts showing, the chances are now 13/44.
**2.** It is more likely that a heart will be dealt now.

# G: Answers to Exercise Sets

## Answers to Exercise Set 1.2

| | | | |
|---|---|---|---|
| **1.** F | | **41.** 185 |
| **2.** F | | **42.** 245 |
| **3.** T | | **43.** 343 |
| **4.** F | | **44.** 533 |
| **5.** T | | **45.** 480 |
| **6.** T | | **46.** 674 |
| **7.** F | | **47.** 944 |
| **8.** T | | **48.** 1271 |
| **9.** T | | **49.** 1473 |
| **10.** F | | **50.** 2192 |
| **11.** 66 | | **51.** 18 |
| **12.** 92 | | **52.** 38 |
| **13.** 76 | | **53.** 47 |
| **14.** 103 | | **54.** 58 |
| **15.** 166 | | **55.** 42 |
| **16.** 185 | | **56.** 54 |
| **17.** 235 | | **57.** 83 |
| **18.** 363 | | **58.** 49 |
| **19.** 533 | | **59.** 32 |
| **20.** 460 | | **60.** 81 |
| **21.** 674 | | **61.** 81 |
| **22.** 964 | | **62.** 102 |
| **23.** 1241 | | **63.** 106 |
| **24.** 1473 | | **64.** 133 |
| **25.** 2251 | | **65.** 154 |
| **26.** 15 | | **66.** 185 |
| **27.** 18 | | **67.** 215 |
| **28.** 19 | | **68.** 343 |
| **29.** 28 | | **69.** 420 |
| **30.** 33 | | **70.** 591 |
| **31.** 41 | | **71.** 694 |
| **32.** 73 | | **72.** 914 |
| **33.** 37 | | **73.** 1251 |
| **34.** 32 | | **74.** 1473 |
| **35.** 61 | | **75.** 2281 |
| **36.** 79 | | **76.** 389 |
| **37.** 92 | | **77.** 4,968 |
| **38.** 96 | | **78.** 3,482 |
| **39.** 113 | | **79.** 41,482 |
| **40.** 176 | | **80.** 448 |

## Answers to Exercise Set 1.3

1. 67
2. 43
3. 333
4. 43
5. 0
6. 28
7. 55
8. 40
9. 4
10. 191
11. 932,746
12. 328,940,911
13. 175,284,131
14. 108,280
15. 387,197,899,079

## Answers to Exercise Set 1.4

1. $47 \times 32$
2. $43 \times 123$
3. $52 \times 182$
4. $21 \times 217$
5. $24 \times 136$
6. 13,902,224
7. 78,942,384
8. 140,544
9. 315,068
10. 592,480
11. 4,399
12. 13,225
13. 5,524,582
14. 7,569
15. 4,416
16. 24,600
17. 980
18. 10
19. 5,000
20. 800
21. $700 \times 70 = 49,000$
22. $500 \times 30 = 15,000$
23. $4,000 \times 240 = 960,000$
24. $5,500 \times 380 = 2,090,000$
25. $3,000 \times 830 = 2,490,000$

## Answers to Exercise Set 1.5

1. 382
2. 384
3. 405 R27
4. 534
5. 645
6. 843
7. 917 R19
8. 903
9. 593
10. 1,596 R7
11. 578
12. 327 R45
13. 931
14. 732 R64
15. 981
16. 890
17. 619 R24
18. 779
19. 910
20. 732 R30
21. 839
22. 837

## Answers to Exercise Set 2.1

1. $\dfrac{42}{112}$
2. $\dfrac{15}{21}$
3. $\dfrac{10}{55}$
4. $\dfrac{102}{114}$
5. $\dfrac{820}{860}$
6. $\dfrac{105}{2,590}$
7. $\dfrac{152}{160}$
8. $\dfrac{135}{330}$
9. $\dfrac{781}{891}$
10. $\dfrac{445}{450}$

## Answers to Exercise Set 2.2

1. composite
2. prime
3. composite
4. prime
5. composite
6. 2(2)(3)
7. 3(3)(7)
8. 5(17)
9. 37
10. 2(2)(2)(2)(2)(3)

## Answers to Exercise Set 2.3

1. 4
2. 12
3. 7
4. 3
5. 16
6. 12
7. 9
8. 18
9. 33
10. 32

11. 8
12. 9
13. 24
14. 16
15. 3
16. $\frac{2}{3}$
17. $\frac{2}{5}$
18. $\frac{2}{5}$
19. $\frac{3}{5}$
20. $\frac{3}{8}$
21. $\frac{4}{11}$
22. $\frac{5}{6}$
23. $\frac{1}{4}$
24. $\frac{5}{8}$
25. $\frac{9}{28}$
26. $\frac{5}{9}$
27. $\frac{1}{3}$
28. $\frac{5}{6}$
29. $\frac{36}{61}$
30. $\frac{1}{7}$

31. $\frac{2}{8}$
32. $\frac{4}{12}$
33. $\frac{8}{20}$
34. $\frac{36}{81}$
35. $\frac{24}{48}$
36. $\frac{14}{49}$
37. $\frac{8}{64}$
38. $\frac{30}{78}$
39. $\frac{9}{24}$
40. $\frac{27}{45}$
41. $\frac{8}{36}$
42. $\frac{44}{60}$
43. $\frac{42}{75}$
44. $\frac{72}{88}$
45. $\frac{40}{96}$
46. $\frac{28}{68}$

**Answers to Exercise Set 2.4**

1. $2\frac{2}{5}$
2. 2
3. $1\frac{7}{12}$
4. $8\frac{3}{5}$
5. $6\frac{1}{2}$
6. $1\frac{1}{2}$

7. $2\frac{2}{7}$
8. $4\frac{2}{3}$
9. $4\frac{3}{4}$
10. 3
11. 2
12. 16
13. $\frac{11}{4}$
14. $\frac{13}{4}$
15. $\frac{24}{5}$
16. $\frac{28}{5}$
17. $\frac{38}{3}$
18. $\frac{75}{4}$
19. $\frac{139}{7}$
20. $\frac{97}{6}$
21. $\frac{86}{7}$
22. $\frac{94}{7}$

23. $\frac{71}{5}$
24. $\frac{112}{5}$
25. $\frac{5}{7}$
26. $1\frac{1}{6}$
27. $3\frac{1}{15}$
28. $\frac{33}{45}$
29. $7\frac{1}{2}$
30. $\frac{7}{8}$
31. $4\frac{19}{22}$
32. $\frac{3}{16}$
33. $\frac{1}{2}$
34. $\frac{5}{9}$
35. $\frac{7}{13}$

**Answers to Exercise Set 2.5**

1. $1\frac{5}{8}$
2. $\frac{2}{9}$
3. $\frac{11}{40}$
4. $1\frac{13}{18}$
5. $\frac{1}{2}$
6. $\frac{1}{8}$
7. $9\frac{1}{4}$
8. $7\frac{3}{4}$
9. $23\frac{17}{18}$
10. $6\frac{13}{24}$
11. $1\frac{7}{12}$
12. $6\frac{13}{24}$

## Answers to Exercise Set 2.6

1. $\dfrac{9}{35}$

2. $\dfrac{1}{4}$

3. $\dfrac{1}{39}$

4. $\dfrac{3}{10}$

5. $4\dfrac{1}{2}$

6. $\dfrac{7}{96}$

7. $\dfrac{25}{54}$

8. $\dfrac{1}{6}$

9. 36

10. 9

11. $1\dfrac{1}{3}$

12. $4\dfrac{4}{7}$

13. $1\dfrac{1}{4}$

14. $4\dfrac{3}{8}$

15. 14

16. 6

## Answers to Exercise Set 2.7

1. $6\dfrac{4}{5}$

2. 40

3. $1\dfrac{1}{4}$

4. $\dfrac{4}{9}$

5. 28

6. $2\dfrac{1}{5}$

7. $46\dfrac{7}{8}$

8. $88\dfrac{1}{5}$

9. $80\dfrac{1}{4}$

10. $71\dfrac{1}{9}$

11. $94\dfrac{6}{7}$

12. $37\dfrac{8}{11}$

13. $141\dfrac{1}{2}$

14. $41\dfrac{7}{8}$

15. $57\dfrac{3}{8}$

16. 7,650

17. $2\dfrac{18}{35}$

18. $2\dfrac{17}{56}$

19. $3\dfrac{11}{108}$

20. $2\dfrac{91}{110}$

21. $6\dfrac{17}{30}$

22. $3\dfrac{51}{79}$

23. $6\dfrac{162}{167}$

24. $23\dfrac{9}{11}$

25. $5\dfrac{51}{148}$

26. $7\dfrac{14}{17}$

## Answers to Exercise Set 3.1

1. Two hundred sixty-five thousandths
2. Seventy-nine hundredths
3. Eight hundred forty-two thousandths
4. Three thousand nine hundred eleven ten thousandths
5. Five thousand seventeen ten thousandths
6. Fifty-three thousandths
7. Sixty-one hundred thousandths
8. Two thousand one ten thousandths
9. Forty-five hundredths
10. One ten thousandths
11. .250
12. 4.23
13. 12.040
14. .004062
15. 715.8
16. .300
17. 19,000.0037
18. .80
19. .0016
20. .051

## Answers to Exercise Set 3.2

1. .3
2. .05
3. .321
4. 12.01
5. 124.0003
6. 18.7
7. .300
8. 1.45
9. 22.3
10. 4.330
11. .5
12. .75
13. .375
14. .3125
15. .5625
16. .53125
17. .875
18. .875
19. .6875
20. .875

## Answers to Exercise Set 3.3

1. $\dfrac{1}{100}$

2. $\dfrac{1}{2}$

3. $\dfrac{5}{8}$

4. $2\dfrac{1}{10}$

5. $23\dfrac{9}{20}$

6. $\dfrac{1}{1,250}$

7. $\dfrac{38}{625}$

8. $\dfrac{2,341}{10,000}$

9. $\dfrac{4,329}{100,000}$

10. $18\dfrac{1}{50}$

## Answers to Exercise Set 3.4

1. .77
2. 25.03
3. 641.099
4. 22.165
5. 38.89
6. 12.42
7. .0045
8. .869
9. .802
10. .08
11. 63.554
12. 10.439
13. 51.292
14. 44.3456
15. 57.3583
16. 4.51235

## Answers to Exercise Set 3.5

1. 74
2. 9.36
3. 32.40
4. 30.602
5. .4144
6. .067648
7. .98076
8. .004284
9. .001803
10. .0000896
11. 32.67
12. .3267
13. 2.86268
14. .4077
15. 6.0088
16. 87
17. .069
18. 9560
19. 4.53
20. 4069
21. .94
22. 92
23. 749
24. 5.3479
25. .492568
26. .0249653
27. .05908
28. .00007156
29. .495674
30. .00000038649

## Answers to Exercise Set 3.6

1. .170
2. .050
3. .600
4. 4.730
5. 420.000
6. 3.700
7. .047
8. .670
9. 404.286
10. 36.818

## Answers to Exercise Set 4.1

1. 83%
2. 44%
3. 92%
4. 9%
5. 61%
6. 40%
7. 55%
8. 38%
9. 39%
10. 252%
11. 396%
12. 98%
13. 7
14. 7%
15. 21%

## Answers to Exercise Set 4.2

1. $\dfrac{1}{100}$

2. $\dfrac{1}{50}$

3. $\dfrac{1}{25}$

4. $\dfrac{7}{100}$

5. $\dfrac{1}{200}$

6. $\dfrac{1}{16}$

7. $\dfrac{1}{15}$

8. $\dfrac{3}{40}$

9. $\dfrac{1}{300}$

10. $\dfrac{3}{400}$

11. $\dfrac{3}{200}$

12. $\dfrac{7}{200}$

13. .42
14. .165
15. 2.31
16. .01
17. .0025
18. .11
19. .38
20. .67
21. .04
22. .44
23. .215
24. .50375

## Answers to Exercise Set 4.3

1. $43\frac{3}{5}\%$
2. 21%
3. 432%
4. 99%
5. 74%
6. 215%
7. 11%
8. 68%
9. $87\frac{1}{2}\%$
10. 75%
11. $32\frac{1}{2}\%$
12. $567\frac{1}{2}\%$
13. $1{,}989\frac{1}{2}\%$
14. $253\frac{1}{10}\%$
15. $5{,}024\frac{9}{10}\%$

## Answers to Exercise Set 4.4

1. 186
2. 12
3. 10
4. $50
5. $\frac{19}{20}$
6. 25%
7. 50
8. 7%
9. 50%
10. 75%
11. 48
12. 50
13. 320
14. 800
15. 96

## Answers to Exercise Set 4.5

1. $\frac{5}{6}$
2. $\frac{2}{9}$
3. $\frac{1}{6}$
4. $\frac{3}{53}$
5. $\frac{1}{11}$
6. $\frac{7}{12}$
7. $\frac{12}{11}$
8. $\frac{5}{12}$
9. $\frac{11}{200}$
10. $\frac{6}{7}$
11. $\frac{39}{2}$
12. Ted gets $130, and Tasha gets $195
13. 6 eggs
14. 5.76 ounces
15. First basket has 24 apples, and second basket has 12 apples
16. 6
17. 12
18. 6
19. 2
20. 21
21. 4
22. 3
23. 125
24. 3
25. 18
26. 60 days
27. $45\frac{1}{2}$ mph
28. 880 revolutions

## Answers to Exercise Set 5.2

1. 23
2. −36
3. −7
4. −3
5. 28
6. −9
7. −24
8. 66
9. 130
10. 3
11. −3
12. −25
13. 280
14. −141
15. −160

## Answers to Exercise Set 5.3

1. −32
2. 216
3. 72
4. −144
5. −3
6. 4
7. $2\frac{4}{5}$
8. −12
9. −64
10. 4
11. $-21\frac{1}{3}$
12. $-5\frac{1}{3}$
13. 120

**14.** $-7\frac{1}{2}$

**15.** 21

**16.** $-42$

**17.** 777

**18.** $-60$

**19.** $-\frac{1}{2}$

**20.** 91

## Answers to Exercise Set 5.4

**1.** 3
**2.** $-8$
**3.** 108
**4.** 539
**5.** 8.4
**6.** 132
**7.** 1,870
**8.** 98
**9.** 165
**10.** 282.2

**11.** 23
**12.** 110
**13.** 13
**14.** 20
**15.** 12
**16.** 44
**17.** 3
**18.** 55
**19.** 1,260
**20.** 3

## Answers to Exercise Set 5.5

**1.** 57
**2.** 275
**3.** 64
**4.** 24
**5.** 37

**6.** 191
**7.** 458
**8.** 30
**9.** 118
**10.** 56

## Answers to Exercise Set 6.1

**1.** $p = 2l + 2w$
**2.** $d = rt$
**3.** $H = \dfrac{av}{746}$
**4.** $I = PRT$
**5.** $A = \dfrac{W}{V}$

**6.** $P = M - O$
**7.** $d = 16t^2$
**8.** $A = S^2$
**9.** $C = \dfrac{5}{9}(F - 32°)$
**10.** $R = \dfrac{N}{T}$

## Answers to Exercise Set 6.2

**1.** $-20d$
**2.** $4b$
**3.** $12x$
**4.** $5a + 2b$
**5.** $15a + 3b - 5$
**6.** $-4ab$
**7.** $18x - 9t$
**8.** $-86y + x$
**9.** $12x - 54r$
**10.** $-34t - 22y$
**11.** $909t$
**12.** $45 - 21y - 11t$
**13.** $99x + 26y$
**14.** $90x - 130y$
**15.** $392r - 17y$

## Answers to Exercise Set 6.3

**1.** $x + y = 5$
**2.** $x + y + z = 9$
**3.** $2x + 2y = 10$
**4.** $x + y - z = 1$
**5.** $x^2 + y^2 = 13$
**6.** $y - 3 = 0$
**7.** $2xz = 16$
**8.** $a^5$
**9.** $4^3 d^3$
**10.** $3^3 y^2 t^2$
**11.** $7^3$
**12.** $3^2 b^2$

**13.** 36
**14.** 729
**15.** $\dfrac{1}{64}$
**16.** 186,624
**17.** 32,768
**18.** 43,000,000
**19.** 620,000
**20.** 0.00000028
**21.** 0.0025
**22.** 122,000,000

## Answers to Exercise Set 6.4

**1.** $176x^2 y + 14y$
**2.** $33x - 71y$
**3.** $-12y$
**4.** $23x - 9$
**5.** $9x$
**6.** $143xy - 5x$
**7.** $780rt + 23t$
**8.** $1,102w - 9$
**9.** $(55x^2)/t$
**10.** 3
**11.** $\dfrac{5xy}{4ab}$
**12.** $\dfrac{a^3}{a - x}$
**13.** $\dfrac{x - y + c}{x - y}$
**14.** $\dfrac{a(x + a)}{ab}, \dfrac{a^2}{ab}, \dfrac{b(a - x)}{ab}$
**15.** $\dfrac{4x(x - 2)}{3}$
**16.** $\dfrac{3}{4}$
**17.** $x$
**18.** $\dfrac{4x^2 - 5x + 3}{(x - 1)^3}$
**19.** $2a + \dfrac{3(a - b)}{c}$
**20.** $\dfrac{a^2 + x^2}{a^2 - x^2}$

## Answers to Exercise Set 6.5

**1.** $p = 5$

**2.** $n = 12\frac{1}{2}$

**3.** $x = 28$

**4.** $c = 6$

**5.** $y = 4$

**6.** $n = 36$

**7.** $a = 48$

**8.** $b = Wc$

**9.** $A = \dfrac{W}{V}$

**10.** $W = \dfrac{P}{AH}$

## Answers to Exercise Set 7.1

**1.** 8

**2.** 10

**3.** 9

**4.** 3

**5.** 5

**6.** 12

**7.** 10

**8.** 1

**9.** .2

**10.** .3

**11.** 73

**12.** 35

**13.** 54.2

**14.** 17.68

**15.** 20.69

**16.** 26

**17.** 43

**18.** 56

**19.** 85

**20.** 97

**21.** 135

**22.** 223

**23.** 343

**24.** 739

**25.** 487

**26.** $\dfrac{10}{12}$

**27.** $\dfrac{14}{5}$

**28.** $\dfrac{124}{49}$

**29.** $\dfrac{65}{20}$

**30.** $\dfrac{1,068}{400}$

**31.** $\dfrac{5}{6}$

**32.** $\dfrac{31}{36}$

**33.** $\dfrac{63}{72}$

**34.** $\dfrac{7}{4}$

**35.** $\dfrac{6}{8}$

**36.** 9

**37.** 7

**38.** $2\frac{1}{3}$

**39.** $3\frac{1}{5}$

**40.** $1\frac{1}{2}$

## Answers to Exercise Set 7.2

**1.** $7abc(ac^2 - 4)$

**2.** $ax^2(9a - 16)$

**3.** $5x(1 - 5x)$

**4.** $(7x^2 - 4y)(7x^2 + 4y)$

**5.** $4a^3b(4ab^2 + 1)$

**6.** $11x^2(3xy^5 - z)$

**7.** $7x^2y^2(3y^2 - xy)$

**8.** $x^2(x - 8)$

**9.** $a(a + b^3)$

**10.** $3a(2a^4 - b^2)$

## Answers to Exercise Set 7.3

**1.** $(a^2 + a + 1)(a^2 - a + 1)$

**2.** $(x + 3)(x + 7)$

**3.** $(x - 15)(x - 3)$

**4.** $(x + 9)(x - 4)$

**5.** $(x - 16)(x + 3)$

**6.** $(x - 3y)(x - 11y)$

**7.** $(x + b)^2(x - b)^2$

**8.** $(x - 6)(x - 6) = (x - 6)^2$

**9.** $(z^2 + 8)^2$

**10.** $(b + c)^2$

## Answers to Exercise Set 7.4

**1.** $5ad(3ac + 4a - 3cd)$

**2.** $(2x + 3y)^2$

**3.** $(3ab - 4ac)^2$

**4.** $(3x - y + 2z)(x + 3y)$

**5.** $(3x + 9)(2x + 1)$

**6.** $(5x - 7)(3x + 3)$

**7.** $(4x + 13)(3x - 3)$

**8.** $(3a - 4b)^2$

**9.** $(3x^2 - 2)^2$

**10.** $(2x - 5y)^2$

**11.** $\dfrac{x + a}{3(x - a)}$

**12.** $x - a + \dfrac{3}{x - a}$

**13.** $\dfrac{x(1 - x)^2}{(1 - x)^3}, \dfrac{x^2(1 - x)}{(1 - x)^3},$ $\dfrac{x^3}{(1 - x)^3}$

**14.** $\dfrac{2(x + y)}{a}$

**15.** $x^2 - y^2$

**Answers to Exercise Set 8.1**

1. $(2,2)$
2. $(4,-1)$
3. $(3,1)$

4. $(-2,3)$
5. $(3,-2)$
6. $(-2,-3)$

7.

9.

8.

10.

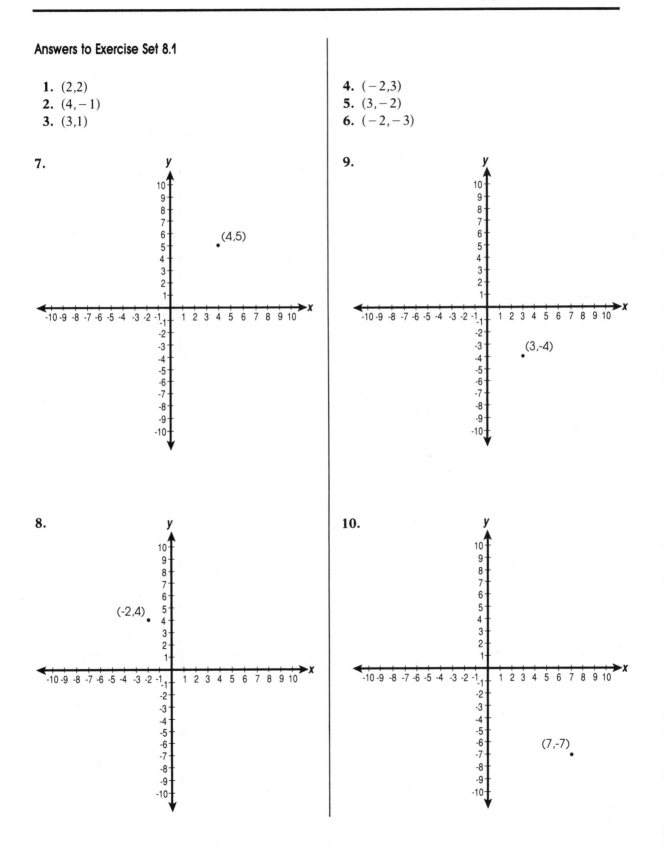

**Answers to Exercise Set 8.2**

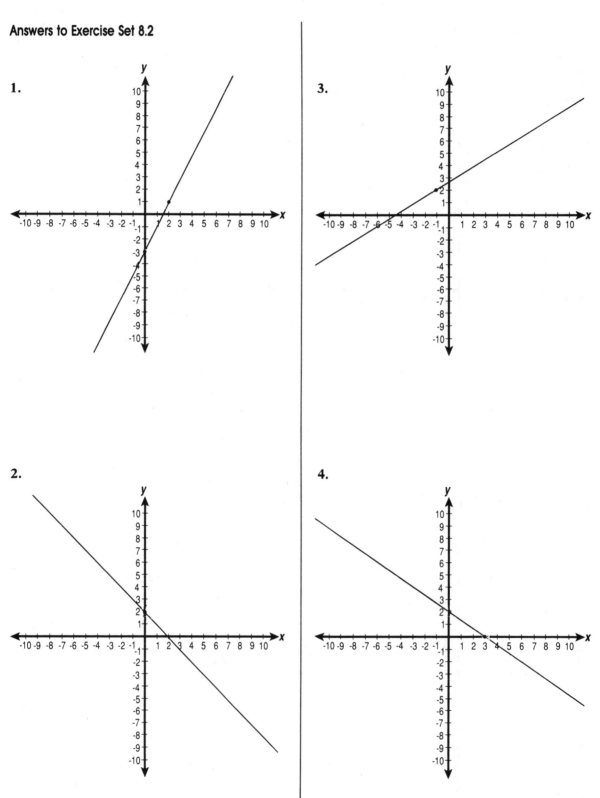

1.

2.

3.

4.

**5.**

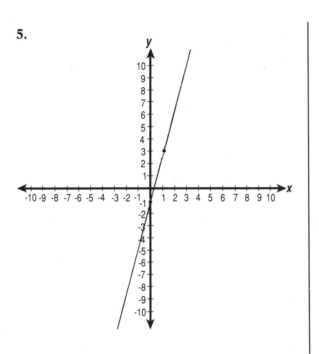

### Answers to Exercise Set 8.2

**1.** $m = 2$

**2.** $m = -\dfrac{1}{3}$

**3.** $m = -2$
**4.** $m = 1$
**5.** $m = 5$
**6.** $m = -1$
**7.** $m = 0$
**8.** No slope
**9.** No slope
**10.** $m = 0$

### Answers to Exercise Set 8.5

**1.** $x = 3, y = 3$
**2.** $x = 0, y = 2$
**3.** infinite number of solutions
**4.** no solutions
**5.** no solutions
**6.** $x = 3, y = 2$

**7.** Let $x$ = no. of min. spaces passed over by min. hand; $y$ = no. passed by hr. hand. $12y = x$; $y = x - 60$. Subtracting, $11y = x - (x - 60) = 60$; $y = 5\dfrac{5}{11}$. $5\dfrac{5}{11}$ min. spaces past 12 o'clock gives $1.05\dfrac{5}{11}$ o'clock.

**8.** $a$ = first; $b$ = second. $a + \dfrac{b}{2} = 35$; $\dfrac{a}{2} + b = 40$. Multiply first equation by 2, $2a + b = 70$. Subtracting second equation from this $1\dfrac{1}{2}a = 30$; $a = 20$; $a + \dfrac{b}{2} = 35$; $\dfrac{b}{2} = 35 - 20$; $\dfrac{b}{2} = 15$; $b = 30$.

**9.** $a$ = Janet's profit; $b$ = Tim's profit. $a + b = \$153$; $a - b = \$45$. Adding, $2a = \$198$; $a = \$99$; $b = 54$. Dividing $918 in the proportions of 99 and 54, $918 \div 153 = 6$; $99 \times 6 = \$594$ for Janet; $54 \times 6 = \$324$ for Tim.

### Answers to Exercise Set 8.6

**1.** $x = 3, y = 2$
**2.** $x = -5, y = 10$
**3.** $x = -3, y = 3$
**4.** $x = 1, y = 3$
**5.** $x = 18, y = -2$

**6.** Try elimination by substitution. $x$ = part inv. at 5%; $y$ = part inv. at 6%. $.05x + .06y = \$1,220$; $x + y = \$22,000$. Multiplying $.05x$, etc., by 20 we get $x + 1.20y = \$24,400$, from which $x = \$24,400 - 1.20y$. Substituting this value in other equation, $\$24,400 - 1.20y + y = \$22,000$; $-.20y = -\$2,400$; $y = \$12,000$; hence $x = \$10,000$.

**7.** Try elimination by comparison. $a$ = Jack's age; $b$ = Joe's age. $a = 2b$; $a - 20 = 4(b - 20)$; $a = 4(b - 20) + 20$; hence $2b = 4b - 80 + 20$; $2b - 4b = -60$; $b = 30$ years for Joe; $2 \times 30 = 60$ years for Jack.

**8.** $a + \dfrac{b}{3} = \$1,700$; $\dfrac{a}{4} + b = \$1,800$. Multiplying first equation by 3, $3a + b = \$5,100$. Subtracting second equation, $2\dfrac{3}{4}a = \$3,300$; $a = \$1,200$. Substituting in first equation $\$1,200 + \dfrac{b}{3} = \$1,700$; $\dfrac{b}{3} = \$1,700 - \$1,200 = \$500$; $b = \$1,500$.

**9.** $\dfrac{a}{2} + \dfrac{b}{3} = 45$; $\dfrac{a}{5} + \dfrac{b}{2} = 40$. Multiplying both equations, $a + \dfrac{2b}{3} = 90$; $a + \dfrac{5b}{2} = 200$. Subtracting the first from the second $\dfrac{11b}{6} = 110$; $b = 60$. Substituting in first equation, $\dfrac{a}{2} + 20 = 45$; $\dfrac{a}{2} = 25$; $a = 50$.

**10.** Try substitution. $14A + 15M = \$153$; $6A - 4M = \$3$; $6A = \$3 + 4M$, whence $A = \$.50 + \dfrac{2M}{3}$. Substituting in other equation $\$7 + \dfrac{28M}{3} + 15M = \$153$; $9\dfrac{1}{3}M + 15M = \$153 - \$7$. $\dfrac{73M}{3} = \$146$; $M = \$6$. Substituting in original equation, $6A - \$24 = \$3$; $6A = \$27$; $A = \$4.50$.

**11.** There are several ways to solve problems like this. The method by simultaneous equations might be as follows. Select letters to represent values that do not change. $a$ = wt. of copper to be added; $b$ = wt. of tin, $b = 80 - \dfrac{7b}{3} = 24$. $a = \dfrac{11b}{4} - \dfrac{7b}{3} = 66 - 56 = 10$ lbs.

**12.** Eliminate by comparison. $B + \dfrac{J}{8} = \$1,200$; $\dfrac{B}{9} + J = \$2,500$. $B = \$1,200 - \dfrac{J}{8}$; $B = \$22,500 - 9J$; $\$1,200 - \dfrac{J}{8} = \$22,500 - 9J$; $9J - \dfrac{J}{8} = \$22,500 - \$1,200$; $\dfrac{71J}{8} = \$21,300$; $J = \$2,400$; $B + \dfrac{J}{8} = \$1,200$; $B + \$300 = \$1,200$; $B = \$900$.

## Answers to Exercise Set 9.1

1. ascending
2. descending
3. ascending
4. descending
5. ascending
6. ascending
7. ascending
8. descending
9. descending
10. ascending

## Answers to Exercise Set 9.2

1. 12, 16
2. 25, 31
3. 0, 12
4. 20, 11
5. 15, 30
6. 55, 22
7. 16, 24
8. 35, 20
9. 27, 9
10. 28, 35

## Answers to Exercise Set 9.3

1. 128, 256
2. 12, 6
3. 1,250, 6,250
4. 27, 1
5. 192, 12,288
6. 16, 256
7. 81, 243
8. $5^3$, $5^4$
9. 384, 1,536
10. 100, 500

## Answers to Exercise Set 9.4

1. 56
2. 252
3. 1,300
4. 254
5. 9,840
6. 19,530
7. 100
8. 91
9. 10,366
10. 3,124

## Answers to Exercise Set 10.2

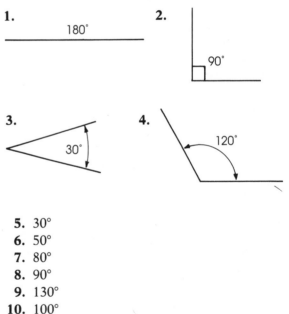

5. 30°
6. 50°
7. 80°
8. 90°
9. 130°
10. 100°
11. 50°

## Answers to Exercise Set 10.3

1. ∠2 = 30°. Angles that coincide are equal.
2. ∠ABD = 22° 30′. A bisector divides an angle in half.

**3.** (a) $\angle 1 = \angle 3$ ⎫
    (b) $\angle 2 = \angle 5$ ⎬ Axiom 1
    (c) Relationship unknown

**4.** $\angle 1$ and $\angle 2$, $\angle 2$ and $\angle 3$, $\angle 3$ and $\angle 4$, $\angle 4$ and $\angle 5$, $\angle 5$ and $\angle 6$, $\angle 6$ and $\angle 7$, $\angle 7$ and $\angle 1$

**5.** $\angle 1$ and $\angle 5$, $\angle 2$ and $\angle 6$

**6.** $\angle 2 = 50°$, $\angle 4 = 30°$, $\angle 5 = 50°$, $\angle 6 = 100°$

**7.** $\angle AOC = 80°$

**8.** $\angle AOD = 180°$

**9.** $\angle BOE = 180°$

**10.** $\angle FOB = 130°$

**11.** $67° \ 30'$

**12.** $60°$

**13.** $45°$

**14.** $30°$

**15.** $22° \ 30'$

**16.** $22°$

**17.** $45°$

**18.** $35°$

**19.** $58°$

**20.** $85°$

**21.** $56° \ 30'$

**22.** $155°$

**23.** $55°$

**24.** $136°$

**25.** $92°$

**26.** $105° \ 30'$

**27.** $101° \ 30'$

**28.** alternate interior

**29.** alternate interior

**30.** corresponding

**31.** alternate exterior

**32.** corresponding

**33.** $\angle 1 = 50°$, $\angle 2 = 130°$, $\angle 4 = 130°$

**34.** $\angle 6 = 140°$, $\angle 7 = 140°$, $\angle 8 = 40°$

**35.** Two lines perpendicular to a third line are parallel.

**36.** (a) alternate interior angles are equal

(b) corresponding angles are equal

(c) alternate exterior angles are equal

**37.** If a pair of alternate interior angles are equal, the lines are parallel.

**38.** $\angle 3$ is supplementary to $115°$. Therefore $\angle 3$ equals $65°$, making corresponding angles equal.

**39.** $70°$

**40.** Extend $AB$ to $D$ and construct $\angle BDE$ equal to $60°$. Then $DE$ is parallel to $BC$ because corresponding angles are equal.

**41.** If the triangle is moved along the edge of the T-square into any two different positions, then lines drawn along side $a$ will be parallel to each other, and lines drawn along side $b$ will also be parallel to each other.

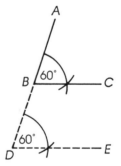

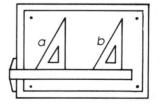

## Answers to Exercise Set 10.4

1. 45°
2. 180°
3. 120°
4. 20° and 70°
5. 52°
6. $32\frac{8}{11}°$, $49\frac{1}{11}°$, $98\frac{2}{11}°$ (Let $x$ = angle; then $x + \frac{1}{2}x + \frac{1}{3}x = 180°$ and $x = 98\frac{2}{11}°$.)
7. Ratio is 1:1 or equal
8. $\angle 2$ is supplementary to $\angle 1$ and $\angle 3$ is supplementary to $\angle 4$.
Therefore $\angle 2 = \angle 3$
Therefore $AB = BC$ and $\triangle ABC$ is isosceles
9. In $\triangle ABC$ $\angle 1 = \angle 2$
$\angle 1 = \angle 3$
Therefore $\angle 3 = \angle 2$
Therefore $DA = DE$
10. $AB = AD$, $\angle 1 = \angle 2$, and $AC = AC$
Therefore $\triangle ABC$ is congruent to $\triangle ADC$ by SAS
11. $AD = DC$, $BD = BD$, and $\angle ADB = \angle CDB$
Therefore $\triangle ABD$ is congruent to $\triangle CBD$ by SAS
12. $\angle 3 = \angle 5$, $BC = CD$, and $\angle 2 = \angle 6$
Therefore $\triangle ABC$ is congruent to $\triangle EDC$ by ASA
13. $AB = BD$, $EB = BC$, and $\angle 1 = \angle 2$
$\triangle ABE$ is congruent to $\triangle CBD$ by SAS
$\angle 3 = \angle 4$
Therefore $AE$ is parallel to $CD$ (two lines are parallel if a pair of alternate interior angles are equal)
14. $AD = BC$, $AC = BD$, $AB = AB$

Therefore $\triangle BAD$ is congruent to $\triangle ABC$
Therefore $\angle 1 = \angle 2$.
15. $AB = CB$, $AD = CD$
$DB = DB$
Therefore $\triangle ABD$ is congruent to $\triangle CBD$ by SSS.
Therefore $\angle 5 = \angle 6$
Therefore $\angle 7 = \angle 8$
$DE = DE$
Therefore $\triangle ADE$ is congruent to $\triangle CDE$ by SAS.
Therefore $\angle 1 = \angle 2$
16. $AB = EF$, $\angle A = \angle F$, and $\angle C = \angle D$
Therefore $\triangle ABC$ is congruent to $\triangle DEF$ by SAA
Therefore $BC = DE$

## Answers to Exercise Set 10.5

1. 18 inches
2. 615.44 inches
3. 452.16 inches
4. 43.96 inches
5. 157 feet

## Answers to Exercise Set 10.6

1. parallelogram
2. square
3. rhombus
4. trapezoid
5. rectangle
6. trapezium
7. 22 feet
8. 24 feet
9. 28 feet
10. 1,880 feet

## Answers to Exercise Set 10.7

1. Yes
2. No
3. No
4. No
5. No
6. Yes
7. No
8. No
9. No
10. No

## Answers to Exercise Set 11.1

1. 21 square centimeters
2. 3,500 square feet
3. 9 square feet
4. 15,600 square feet
5. $1,125
6. 64 flagstones
7. 110 feet
8. 24 feet long
9. $20.16
10. 675 square feet
11. 336 square feet
12. 120 square feet

## Answers to Exercise Set 11.2

1. $5,626.88
2. 4 inches
3. 4 times as great
4. 28 feet
5. $615.44
6. 615.44 square inches of pizza
7. 147 square feet
8. 17 yards
9. 49 square inches
10. 4 feet

## Answers to Exercise Set 11.3

1. 22,464 cubic inches
2. 513,216 cubic inches
3. 243,648 cubic inches
4. 0.078125 cubic feet
5. 9 cubic feet
6. 10 cubic yards
7. $411\frac{1}{3}$ cubic yards
8. 46,656 cubic inches
9. 10,368 cubic inches
10. 297 cubic inches
11. 154 square inches
12. 904.32 cubic feet
13. 803.84 square inches
14. $37\frac{1}{3}$ cubic inches
15. 401.92 cubic yards

## Answers to Exercise Set 11.4

1. 3.74623 kilometers
2. 4,253 meters
3. .00008546 square kilometers
4. 473,860 square millimeters
5. 3,560,000 cubic centimeters
6. .000374658 cubic meters
7. 31,237.65 centiliters
8. 3.123765 hectoliters
9. 746,000 centigrams
10. 3.426 kilograms
11. 951.9528 acres
12. 1,064.175 bushels
13. 95.0696 miles
14. 1,276.086 gallons
15. 166.4473 pounds
16. 79.5528 meters
17. 63.6775 hectares
18. 65.4857 dekaliters
19. 65.8801 hectoliters

### Answers to Exercise Set 12.1

1. 6 feet long
2. $1\frac{1}{5}$ inches long
3. 32 feet high
4. $33\frac{1}{3}$ inches long
5. 36 feet long

### Answers to Exercise Set 12.2

1. 130 million
2. 1930
3. 178 million
4. 1950
5. 1929

### Answers to Exercise Set 12.3

1. 19 deaths per 100,000 people
2. 29 deaths per 100,000 people
3. 23 deaths per 100,000 people
4. 27 deaths per 100,000 people
5. 1950
6. 1930
7. 18 miles
8. $6\frac{3}{5}$ hours
9. 6 miles
10. 48 miles
11. 2 hours

### Answers to Exercise Set 12.4

1. 1970
2. 1930
3. 1980

4. 1950 and 1960
5. 8,000 people
6. 200,000 people
7. 600,000 people
8. Public utilities
9. 2,400,000 people
10. Armed forces

### Answers to Exercise Set 12.5

1. $300
2. Rent
3. Overhead
4. Material
5. Labor

6.

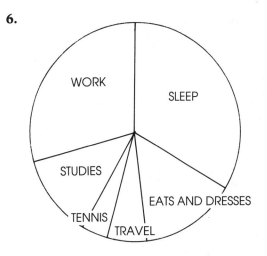

WORK 7/24=29.2%

EATS AND DRESSES 7/48=14.6%

STUDIES 3/24=1/8=12.5%

TRAVEL 3/48=6.3%

TENNIS 1/24=4.2%

SLEEP 8/24=1/3=33.3%

**Answers to Exercise Set 13.1**

1. 24
2. $P(5,5) = 120$
3. $P(12,5) = 95,040$
4. $(8 - 1)! = 720$

5. The two vowels can be arranged from five in any of

$$P(5,2) = \frac{5!}{(5 - 2)!} = \frac{5!}{3!}$$
$$= 5 \cdot 4 = 20 \text{ ways.}$$

The three other letters can be arranged from 21 in

$$P(21,3) = \frac{21!}{(21 - 3)!}$$
$$= 21 \cdot 20 \cdot 19$$
$$= 7,980 \text{ ways.}$$

Hence, all five letters can be arranged in

$$P(5,2)P(21,3) = 20(7,980)$$
$$= 159,600 \text{ ways}$$

6. Here $n_1 = 3$, $n_2 = 6$, $n_3 = 4$, $n = 3 + 6 + 4 = 13$, and

$$P_a = \frac{n!}{n_1!n_2!n_3!}$$
$$= \frac{13!}{3!6!4!}$$
$$= \frac{13 \cdot 12 \cdot 11 \cdot 10 \cdot 9 \cdot 8 \cdot 7}{3 \cdot 2 \cdot 4 \cdot 3 \cdot 2}$$
$$= 60,060$$

7. $P(4,4) = 24$
8. 120
9. 56
10. 7

**Answers to Exercise Set 13.2**

1. $C(4,3) = 4$
2. This is a question of grouping without regard to order, with $n = 12$, and $t = 5$, where $C(12,5) = 792$.
3. $N = 4(6) = 24$
4. $N = 4(3)(6) = 72$
5. $N = 2^5 = 32$
6. $N = 6^4 = 1,296$
7. $N = 2(6)(4)(9) = 432$ models
8. Excluding the one case in which no dot is raised:
$N = 2^6 - 1 = 64 - 1 = 63$
9. Here duplications of flags are possible. With $n = k = 5$, we get
$N = n^k = 5^5 = 3,125$
10. $M = 2^{16} - 1 = 65,536 - 1 = 65,535$

**Answers to Exercise Set 13.3**

1. The required array and corresponding $h$ values are:

$$h_5 = 4 \begin{cases} h_1 = 1 & H_1H_2H_3 \\ h_2 = 3 \begin{cases} H_1H_2T_3 \\ H_1T_2H_3 \\ T_1H_2H_3 \end{cases} \\ h_3 = 3 \begin{cases} H_1T_2T_3 \\ T_1H_2T_3 \\ T_1T_2H_3 \end{cases} \\ h_4 = 1 & T_1T_2T_3 \end{cases}$$

Hence, the corresponding $f$ values are:

$$f_1 = f_4 = 8 - 1 = 7$$
$$f_2 = f_3 = 8 - 3 = 5$$
$$f_5 = 8 - 4 = 4$$

**2.** $p_1 = p_4 = \frac{1}{8} = 12.5\%$
$q_1 = q_4 = \frac{7}{8} = 87.5\%$
$p_2 = p_3 = \frac{3}{8} = 37.5\%$
$q_2 = q_3 = \frac{5}{8} = 62.5\%$
$p_5 = q_5 = \frac{4}{8} = 50.0\%$

**3.** Events $E_1$, $E_2$, $E_3$, and $E_4$ are all the mutually exclusive different outcomes possible when three coins are tossed. Therefore:

$$p_1 + p_2 + p_3 + p_4$$
$$= \frac{1}{8} + \frac{3}{8} + \frac{3}{8} + \frac{1}{8}$$
$$= \frac{8}{8} = 1 \quad \text{Check}$$

Also, event $E_5$ is mutually exclusive with events $E_3$ and $E_4$, to exhaust the same possibilities with a different grouping. Therefore:

$$p_5 + p_3 + p_4 = \frac{4}{8} + \frac{3}{8} + \frac{1}{8}$$
$$= \frac{8}{8} = 1 \quad \text{Check}$$

**4.** Since $E_5$ is the event that either of the mutually exclusive events, $E_1$ or $E_2$, happen

$$p_5 = p_1 + p_2 = \frac{1}{8} + \frac{3}{8} = \frac{4}{8}$$
$$= 0.5 \text{ (as before)}$$

**5.** (a) The odds on $E_1$ happening are:

$h_1{:}f_1 = 1{:}7$, or 1 to 7 for

(b) The odds on $E_2$ happening are:

$f_2{:}h_2 = 5{:}3$, or 5 to 3 against

(c) The odds on $E_5$ happening are:

$h_5{:}f_5 = 4{:}4 = 1{:}1$, or even either way

**6.** (a) For event $E_a$ that the exposed card be a face card, $w = 52$, $h_a = 3 \cdot 4 = 12$, and

$$p_a = h_a/w = 12/52 = 4/13$$

(b) For the event $E_b$ that the exposed card be a black card, $h_b = 2 \cdot 13 = 26$, and

$$p_b = h_b/w = 26/52 = \frac{1}{2}$$

(c) For the event $E_c$ that the exposed card be a black face card, $h_c = 2 \cdot 3 = 6$, and

$$p_c = h_c/w = 6/52 = 3/26$$

(d) Hence, the probability of the partially overlapping event $E_d$, that either $E_a$ or $E_b$ happen, is:

$$p_d = p_a + p_b - p_c$$
$$= \frac{12 + 26 - 6}{52}$$
$$= \frac{32}{52} = \frac{8}{13}$$

**7.** The coins can land in $w = n^k = 2^5 = 32$ ways. Of these, only $h = 2$ (1 all heads, and 1 all tails) comply with the condition that all land the same way. Hence, the required probability is:

$$p = h/w = 2/32 = 1/16$$

Alternatively regarding the required landing as a multiple event, we can reason that the probability of the first coin landing either heads or tails is $p_1 = 2/2 = 1$ (for certainty), but that the probability of the other coins each separately landing the same way is 1/2 in each case.

Hence, $p_2 = p_3 = p_4 = p_5 = 1/2$, and the probability of the specified multiple event is:

$$p = p_1p_2p_3p_4p_5 = 1(\tfrac{1}{2})^4 = 1/2^4$$
$$= 1/16$$

**8.** Any two balls from either box may be combined with any two balls in the other box in

$$w = C_2^{10}C_2^{10} = \left(\frac{10 \cdot 9}{2}\right)^2$$
$$= 45^2 = 2{,}025 \text{ ways}$$

But two red balls from either may be combined with two red balls from the other in only

$$h = C_2^4C_2^4 = \left(\frac{4 \cdot 3}{2}\right)^2$$
$$= 6^2 = 36 \text{ ways}$$

Therefore, the required probability is:

$$p = h/w = 36/2{,}025 = 4/225$$

Alternatively, we may regard the drawing of all four balls as a multiple event consisting of the four separate drawings with the separate probabilities, $p_1 = p_3 = 4/10$ independent of each other, and $p_2 = p_4 = 3/9$ independent of each other but dependent upon $p_1$ and $p_3$ respectively. Then, by the multiplication theorem:

$$p = p_1p_2p_3p_4 = \tfrac{4}{10} \cdot \tfrac{3}{9} \cdot \tfrac{4}{10} \cdot \tfrac{3}{9}$$
$$= (2 \cdot 2)/(5 \cdot 3 \cdot 5 \cdot 3) = 4/225$$

**9.** By the same reasoning, if all four balls are drawn from the same box, then

$$p = h/w = C_4^4/C_4^{10}$$

$$= 1 \left/ \frac{\dfrac{1}{\cancel{3}\ \cancel{2}}{10 \cdot \cancel{9} \cdot \cancel{8} \cdot 7}}{\underset{1\ \ 1\ \ 1}{4 \cdot 3 \cdot 2}} \right.$$

$$= 1/210$$

Or, alternatively, with each separate event dependent upon those preceding it:

$$p = p_1p_2p_3p_4$$
$$= \tfrac{4}{10} \cdot \tfrac{3}{9} \cdot \tfrac{2}{8} \cdot \tfrac{1}{7}$$
$$= \tfrac{1}{210}$$

**10.** Of the 1,098,240 ways of dealing a one-pair hand, 4/13 of these have pairs of jacks or better. We could add this number to all the numbers of still better hands above it in the *table* to find $h =$ the total number of hands which have a pair of jacks or better. However, it is arithmetically simpler to find the number of hands which have single pairs less than jacks, or

$$1{,}098{,}240(9/13) = 760{,}320$$

and add to this the one number of hands in the *table* still lower than these to find,

$$f = 760{,}320 + 1{,}302{,}540$$
$$= 2{,}062{,}860$$

Then, more quickly,

$h = w - f$

$\quad = 2{,}598{,}960 - 2{,}062{,}860$

$\quad = 536{,}100$

and the required probability is

$p = h/w = 536{,}100/2{,}598{,}960$

$\quad = 1$ in 4.8, approximately

**11.** Regardless of the number on which Sloe bets, the mathematical expectation which he obtains for his \$10 is only

$V = A \cdot p = \$360/37 = \$9.73$

In other words, all numbers are equally "unlucky" for the player in the long run. The only reason the "house" wins in the long run is that it never puts *any* money up on its "free" number, zero. Moreover, it cannot afford dishonestly to have the wheel "fixed" to come up more frequently on zero, or on any other number, because—quite aside from any question of ethics or good will—the kind of practiced gambler who frequents such houses would soon detect the trend and "break the bank" by placing large bets on the favored outcomes. In other words, whether Sloe bets with or against it, the house will take his money from him most certainly in the long run by keeping the wheel "honest" and Sloe naively hopeful.

From the theory of probability we learn that, although there may be "systems" which can beat the house that operates a dishonest or accidentally unbalanced wheel, there is no "system" which will long win for anyone but the proprietor on an honest wheel in the long run!

**12.** The given possibilities are the only distributions with less than five cards in any suit. Therefore:

$$p_1 = 0.2155$$
$$p_2 = 0.1054$$
$$\underline{p_3 = 0.0299}$$

$p_1 + p_2 + p_3 = 0.3508 = q$

and the required probability is

$p = 1 - q$

$\quad = 1 - 0.3508 = 0.6492$

**13.** One or more can be drawn in any of

$w = M = n^k - 1 = 2^7 - 1$

$\quad = 128 - 1 = 127$

different combinations. Of these, combinations with the possible even numbers of two, four, or six balls are:

$h = C_2^7 + C_4^7 + C_6^7$

$\quad = \dfrac{7!}{2!5!} + \dfrac{7!}{4!3!} + \dfrac{7!}{6!1!}$

$\quad = \dfrac{7 \cdot 6}{2} + \dfrac{7 \cdot 6 \cdot 5}{3 \cdot 2} + \dfrac{7}{1}$

$\quad = 21 + 35 + 7 = 63$

Hence, the required probability is:

$p = h/w = 63/127$

**14.** The hour hand is between these marks for only 1 hour of 12. Hence, the "continuous" probability of event $E_a$ is:

a) $p_a = 1 \text{ hour}/12 \text{ hours}$

$= 1/12$

However, both hands are between these marks for only 5 minutes out of each 12 hours = 720 minutes. Hence, the "continuous" probability of event $E_b$ is:

b) $p_b = 5 \text{ minutes}/720 \text{ minutes}$

$= 1/144$

Alternatively, we can reason that the minute hand is between these marks for only 5 minutes out of 60, or for 1 hour out of 12. Hence, the probability of the independent separate event $E_m$ that the minute hand stop between the marks is

$p_m = p_a = \frac{1}{12}$

And therefore the probability of the multiple event $E_b$ is:

$p = p_a \ p_m = \frac{1}{12}$

$\frac{1}{12} = \frac{1}{144}$

**Answers to Exercise Set 13.4**

**1.** From the *Mortality Table* we learn that the statistical probability of one such person still being alive at age fifty is

$p_1 = h/w = 69{,}804/100{,}000$

$= 0\ 698 = 69.8\%$

Hence, the probability of the multiple statistical event that two such persons still be alive at age fifty is

$p_2 = p_1$

$p_1 = (0\ 698)^2 = 0.487$

$= 48.7\%$

**Answers to Exercise Set 14.1**

**1.** $\sin B = \dfrac{b}{c}, \cos B = \dfrac{a}{c}, \cot$

$B = \dfrac{a}{b}, \sec B = \dfrac{c}{a}, \csc B = \dfrac{c}{b}$

**2.** $\tan A$

**3.** $\cot A$

**4.** $\sec A$

**5.** $\csc A$

**6.** $\cos A = \frac{4}{5}$

**7.** $\sin A = \frac{3}{5}$

**8.** $\cos A = \frac{15}{17}$

**9.** $\sec A = \frac{17}{15}$

**10.** $\cos A = \frac{12}{13}$

$\tan A = \frac{5}{12}$

$\cot A = \frac{12}{5}$

$\sec A = \frac{13}{12}$

$\csc A = \frac{13}{5}$

## Answers to Exercise Set 14.2

1. cos 64°
2. cot 47°
3. sin 65° 32′
4. tan 1° 10′
5. csc 83° 50′
6. sec $12\frac{1}{2}$°
7. 15° ($90° = 5A + A$; therefore, $90° = 6A$, and $A = 15°$)
8. 45° (reciprocals of the co-functions are equal, therefore, $\angle A = 45°$)
9. 45° ($90° - A = A$; $90° = 2A$; $A = 45°$)
10. 30° ($\cos A = \sin 90° - A$; since $\cos A = \sin 2A$, then $\sin 90° - A = \sin 2A$, $90° - A = 2A$; $3A = 90°$ and $A = 30°$)

## Answers to Exercise Set 14.3

1. .1392
2. .6691
3. .8391
4. .5095
5. 1.079
6. .9063
7. 4.134
8. .9781
9. .3839
10. 6.3925
11. 4695
12. 1.4826
13. .8480
14. .7071
15. .5000
16. 15°
17. 35°
18. 60°
19. 70°
20. 10°

21. $a = 54.46$
22. $a = 3.42$
23. $\angle A = 65° 33′$
24. $c = 29.82$
25. $\angle A = 46° 03′$
26. $\angle A = 53° 8′$
27. $b = 31.86$
28. $b = 62.08$
29. $\angle A = 60°$
30. $c = 42$
31. 36° 52′
32. 64
33. 45.04
34. 18.19

## Answers to Exercise Set 14.4

1. .2672
2. .9013
3. 1.079
4. .7674
5. 1.315
6. sin 5° 10′
7. tan 18° 50′
8. cos 65° 20′
9. cot 49° 40′
10. csc 42° 30′

## Answers to Exercise Set 14.5

1. $a = 7, b = 8.57$
2. $\angle C = 69°, b = 58.91, c = 58.44$
3. $\angle A = 76° 52′, \angle B = 35° 8′, c = 20.95$
4. $\angle A = 51° 24′, \angle B = 48° 49′, \angle C = 79° 47′$
5. $\angle B = 12° 56′, \angle C = 146° 4′, c = 12.43$

# H: Answers to Tests

## Answers to Chapter 1 Test

1. B
2. C
3. A
4. B
5. A
6. C
7. D
8. D
9. C
10. C
11. E
12. B
13. A
14. C
15. B
16. D

## Answers to Chapter 2 Test

1. B
2. D
3. A
4. C
5. E
6. B
7. C
8. B
9. A
10. C
11. C
12. B
13. A
14. B
15. E
16. D
17. B
18. A
19. A
20. B

## Answers to Chapter 3 Test

1. D
2. B
3. D
4. D
5. B
6. C
7. D
8. A
9. C
10. C
11. C
12. E
13. B
14. A
15. C
16. A
17. A
18. C
19. D
20. A
21. D

## Answers to Chapter 4 Test

1. C
2. A
3. D
4. E
5. C
6. C
7. B
8. E
9. C
10. E
11. A
12. D
13. D
14. C

15. B
16. C
17. D
18. B
19. A
20. C
21. A
22. D
23. C
24. B
25. B

## Answers to Chapter 5 Test

1. C
2. B
3. C
4. A
5. B
6. B
7. A
8. B
9. B
10. E
11. C
12. A
13. B
14. E
15. D
16. E
17. D
18. E
19. C
20. E
21. C
22. B
23. E
24. B

## Answers to Chapter 6 Test

1. C
2. B
3. D
4. D
5. A
6. D
7. C
8. D
9. E
10. E
11. C
12. D
13. D
14. B
15. D
16. A
17. D
18. C
19. B
20. A

## Answers to Chapter 7 Test

1. C
2. E
3. B
4. C
5. E
6. B
7. C
8. D
9. B
10. E
11. B
12. C
13. C
14. B

15. D
16. A
17. C
18. C
19. D
20. C

### Answers to Chapter 8 Test

1. C
2. D
3. B
4. C
5. A
6. C
7. C
8. E
9. B
10. C
11. D
12. B
13. D
14. A
15. D
16. D
17. D
18. A
19. B
20. E

### Answers to Chapter 9 Test

1. B
2. D
3. A
4. E
5. E
6. C
7. D
8. E
9. D
10. B
11. E
12. B
13. C
14. E
15. B
16. E
17. B
18. C
19. A
20. D

### Answers to Chapter 10 Test

1. A
2. E
3. C
4. E
5. B
6. C
7. D
8. A
9. B
10. A
11. E
12. A
13. D
14. A
15. B
16. E
17. E
18. C
19. A
20. C
21. C

## Answers to Chapter 11 Test

1. C
2. A
3. B
4. E
5. C
6. C
7. B
8. B
9. E
10. B
11. D
12. B
13. E
14. D
15. C
16. E
17. A
18. B
19. D
20. A
21. B
22. B

## Answers to Chapter 12 Test

1. A
2. E
3. B
4. C
5. B
6. B
7. D
8. B
9. D
10. E
11. D
12. D

13. B
14. C
15. E
16. B
17. E
18. E
19. C
20. A

## Answers to Chapter 13 Test

1. A
2. E
3. A
4. C
5. B
6. B
7. D
8. B
9. B
10. C
11. A
12. B
13. B
14. D
15. B
16. C
17. D
18. A
19. D
20. B

## Answers to Chapter 14 Test

1. C
2. B
3. B
4. A

5. C
6. C
7. B
8. B
9. D
10. C
11. A
12. A
13. C
14. B
15. C
16. B
17. A
18. B
19. C
20. D

**Answers to Final Test**

1. D
2. B
3. D
4. C
5. D
6. B
7. A
8. B
9. C
10. C
11. B
12. C
13. D
14. B

# INDEX